The Biological Weapons Taboo

The Biological Weapons Taboo

Michelle Bentley

OXFORD
UNIVERSITY PRESS

OXFORD
UNIVERSITY PRESS

Great Clarendon Street, Oxford, OX2 6DP,
United Kingdom

Oxford University Press is a department of the University of Oxford.
It furthers the University's objective of excellence in research, scholarship,
and education by publishing worldwide. Oxford is a registered trade mark of
Oxford University Press in the UK and in certain other countries

Published in the United States of America by Oxford University Press
198 Madison Avenue, New York, NY 10016, United States of America

British Library Cataloguing in Publication Data
Data available

Library of Congress Control Number: 2023937470

ISBN 978–0–19–889215–1

DOI: 10.1093/oso/9780198892151.001.0001

Printed and bound in the UK by
Clays Ltd, Elcograf S.p.A.

Acknowledgements

My first thanks go to the Leverhulme Trust. The project was supported by a Research Fellowship grant and this book would not have been possible without the generosity of the Leverhulme and their investment in my research. Thanks also go to all those who helped with the grant application and supported the project, including: Sarah Childs (who has been an amazing mentor to me); Richard Jackson (an inspiration and role model throughout my entire career); David Owen (my former PhD supervisor, who has always been unstinting with his help); and Wyn Bowen (a brilliant scholar who gave his expert advice, both professionally and personally).

My thanks further go to Oxford University Press, especially Dominic Byatt—who has given so much of his time, care, and expertise to this project. I also wish to acknowledge the input of the three anonymous reviewers on the manuscript, who did so much to improve this book. Part of this book was published as a previous article: Michelle Bentley (2022) 'A New Model of "Taboo": Disgust, Stigmatization, and Fetishization', *International Studies Review*: 24(3), viac028: 1–25. My gratitude goes to the editors and anonymous reviewers on this, again for their support, feedback, and encouragement. All the comments from everyone involved in this project's publication have made this book a much stronger piece of work. All mistakes, however, are still my own.

A large number of people have very kindly given me the opportunity to present earlier versions of this research, provided feedback, and helped push the project along—in particular James D. Boys, Brett Edwards, Stuart Gottlieb, Tom Gregory, Pauline Heinrichs, Alexandra Homolar, Peter Jackson, Nicholas Kitchen, Jeff Knopf, Matt Moran, Cheyney Ryan, Scott Sagan, Michal Smetana, and Tom Watts. I would also like to acknowledge Oz Hassan, Adam Quinn, and Oliver Turner for consistently being such wonderful people to work with.

I have worked at the Department of Politics, International Relations, and Philosophy at Royal Holloway for over ten years and have truly excellent colleagues who have supported me throughout my career. I thank all of them immensely for their help, friendship, and their commitment to me and our department. I will not attempt to single out names here, but I hope that those who have been especially important know who they are. For four years I was also an external examiner at University College Cork and have met some superb scholars who have also done much to offer support throughout the writing of this book: Jérôme Aan de Wiel, David Fitzgerald, Detmar Klein, Mervyn O'Driscoll, and David Ryan.

Clare Woodford has combined being a first-class best mate with being an excellent academic collaborator and I could not imagine having written this book without her incredible friendship. Clara Eroukhmanoff has also been a great friend and offered valuable feedback on this book. And Julian Jones is acknowledged here for being a source of wisdom and guidance. He has been an anchor in my life.

Finally, as always, this book is dedicated to my remarkable husband Russell Bentley. This is yet another book that simply would not be here without his selfless, unconditional, and very valued support and care.

Contents

Acronyms

ACDA	Arms Control and Disarmament Agency
ACS	American Chemical Society
AHG	Ad Hoc Group
BWC	Biological Weapons Convention
CC	Chemical Corps
CCD	Conference of the Committee of Disarmament
CIA	Central Intelligence Agency
CWC	Chemical Weapons Convention
CWS	Chemical Warfare Service
DoA	Department of the Army
DoD	Department of Defense
DoS	Department of State
FAS	Federation of American Scientists
FBI	Federal Bureau of Investigation
FSA	Federal Security Agency
HFAC	House Foreign Affairs Committee
ICM	International Congress for Microbiology
IPMG	Interdepartmental Political-Military Group
IR	International Relations
IRC	International Red Cross
ISC	International Scientific Commission
JCS	Joint Chiefs of Staff
NAS	National Academy of Sciences
NPT	Nuclear Non-Proliferation Treaty
NSC	National Security Council
NSSM	National Security Study Memorandum
ORO	Operations Research Office
OSA	Office of Systems Analysis
OSRD	Office of Scientific Research and Development
OST	Outer Space Treaty
PoW	Prisoner of War
PSAC	President's Science Advisory Committee
PSB	Psychological Strategy Board
RevCon	Review Conference
SALT	Strategic Arms Limitations Talks
SFRC	Senate Foreign Relations Committee
SIPRI	Stockholm International Peace Research Institute
UK	United Kingdom
UN	United Nations

UNDC	United Nations Disarmament Commission
UNSC	United Nations Security Council
US	United States of America
USCong	US Congress
USIA	United States Information Agency
USSR	Union of Soviet Socialist Republics
WHO	World Health Organisation
WMD	Weapon(s) of Mass Destruction
WPC	World Peace Council
WRS	War Research Service

Introduction

The First Horseman

He rode in with a voice of shattering thunder—Pestilence,[1] followed by War, Famine, and Death. The Four Horsemen of the Apocalypse. The greatest scourges of humanity. The end of days. Given that these bloodcurdling and macabre figures supposedly embody the absolute worst that humans can suffer, is it then surprising that disease makes the list? Indeed, that sickness constitutes the very first menace out of the gate. Disease is the Damocles' Sword that hangs over us. It is the perpetual fear that our health and bodies are inescapably susceptible to the horror of biological infection. Disease creeps into every aspect of our lives with the odious threat of intense pain, disfigurement, disability, and demise. It is a threat not only to oneself but also to those dearest to us. Disease wades through conflicts, ravages societies, and wipes out entire populations. Simply put, disease is the stuff of nightmares.

Is it then also surprising that bioweapons are considered horrific? Bioweapons are the intentional use of disease to kill, maim, or incapacitate. The World Health Organisation (WHO 2004: 5) defines bioweapons as 'those that achieve their intended target effects through the infectivity of disease-causing microorganisms and other such entities, including viruses, infectious nucleic acids and prions'.[2] Classified as weapons of mass destruction (WMD), biological armaments are set apart as especially terrifying on the basis of the vast destruction they can inflict. Yet the terror that these weapons elicit is also connected to the *type* of destruction caused. Disease comprises a disgusting and shocking contaminant that invades our bodies and destroys us from within. We fear infection and are repulsed by our vulnerability to it. Disease further revolts in that it: is invisible, and that when sickness becomes

[1] Revelation 6 in The Bible (1989) calls the first horseman Conquest—but the figure has come to be named Pestilence in more modern interpretations.

[2] The WHO also quote *The Convention on the Prohibition of the Development, Production and Stockpiling of Bacteriological (Biological) and Toxin Weapons and on Their Destruction* (1972), or the Biological Weapons Convention. The document defines a bioweapon as: 'Microbial or other biological agents, or toxins whatever their origin or method of production, of types and in quantities that have no justification for prophylactic, protective or other peaceful purposes; Weapons, equipment or means of delivery designed to use such agents or toxins for hostile purposes or in armed conflict.'

The Biological Weapons Taboo. Michelle Bentley, Oxford University Press. © Michelle Bentley (2023).
DOI: 10.1093/oso/9780198892151.003.0001

visible this is in the form of often gruesome physical symptoms; takes time to have effect; and is indiscriminate. Our aversion is even more compelling in respect to bioweapons in that these armaments coercively force this repellent contamination upon the body as an act of deliberate harm. Bioweapons are characterized by a 'unique repugnance' and a 'veritable revulsion' that can 'inspire particular human dread' (Westing 1989: 20; Mangold and Goldberg 1999: xi; Enemark 2007: 131). Biowarfare is so horrifying, so immoral, and so exceeds the limits of our toleration that we do not even wish to think of it let alone have anything to do with this vile form of aggression. And we will look down on anyone who would do so as uncivilized and inherently wicked.

This book makes the claim that bioweapons are so horrendous that they are *taboo*. Labelling biowarfare as taboo is not simply a statement that this form of violence is especially abhorrent. The reference to a taboo communicates a deeper claim that the negative perceptions associated with bioweapons dictate how actors understand, construct policy around, and employ these armaments. The taboo exerts a powerful normative prohibition which underpins the rejection of bioweapons as pariah. The taboo delegitimizes these weapons in ways that directly influence actor behaviour. Biological armaments are not just another weapon in our minds, and actors do not treat them as such. The taboo causes those actors to reject bioweapons—and this avoidance shapes all aspects of these armaments' power, possession, proliferation, and use. Without understanding the taboo, therefore, we cannot understand bioweapons. The taboo provides an answer to fundamental questions as to how actors perceive biowarfare and why they do what they do in relation to the threat. This book delivers that necessary understanding by presenting the first comprehensive account of the bioweapons taboo. In doing so, the analysis explains why actors find bioweapons abhorrent, how this sentiment is expressed in the form of political behaviours, how the taboo has influenced past decisions concerning biowarfare, and the taboo's strength of influence.

The need to understand the bioweapons taboo is vital and necessary—as evidenced by the Covid-19 pandemic. Covid-19 exemplifies how pandemics can create mass death as well as crushing social, economic, and political disruption on an international scale. Hot on the heels of the 2014 Ebola outbreak, the horrendous implications of what epidemic and pandemic disease means are now all too tragically apparent. Yet, if disease is harnessed as a weapon, the consequences threaten to be even more catastrophic. Bioweapons potentially pose one of the greatest and most destructive global threats. If the dynamics behind these armaments and their use are not fully comprehended, therefore, such a lack of understanding could prove devastating. Global elites, academics, and the public need to have sufficient understanding of the biowarfare threat if there is any hope that the inherent dangers will not be realized. This book identifies and explains the bioweapons taboo as a core factor in that understanding.

In particular, the book discusses the taboo in terms of why bioweapons are not used. It is generally accepted that biowarfare has not been notably employed within modern conflict (e.g. Spiers 2010: 17). The claim of non-use is difficult for two reasons. First, there *have* been allegations of relatively minor examples of bioweapons employment, although these accusations have rarely been substantiated (Carus 2017a). Second, the concepts of 'weapon' and 'use' are problematic. The definition of a 'bioweapon' is typically comprehended as a military armament that transmits biological harm. The policymakers analysed within this book adopt this conceptualization. Yet the idea of biowarfare can be interpreted in significantly more fluid terms, not least in respect of post-colonial thinking (see David Arnold's (1993) work as a good example). For example, African American populations in the United States of America (US) are shown to have been disproportionately affected by the Covid-19 pandemic due to social policies that have put them at greater risk of infection (Randolph 2021). As a form of racist biopolitics, can this example also be classified as an act of biowarfare? If so, the use of bioweapons is much more pervasive than previously assumed. Yet bioweapons are still seen as fundamentally absent in relation to military strategy and use. In fact, the international community has pushed hard to ensure prohibitive measures which ban biowarfare—including the 1925 Geneva Protocol and the 1972 Biological Weapons Convention (BWC).

W. Seth Carus (2017a: 44) highlights that there currently is 'no clear answer' as to why bioweapons are not employed militarily. Non-use is the 'great mystery of biological warfare' and the question is one of the 'unsolved puzzles in military history' (Regis 1999: 221; Guillemin 2005: 8). In an attempt to provide an explanation, it has been argued that bioweapons lack military utility or are too difficult to deliver—hence the armaments are not employed. Alternatively, it is asserted that actors are deterred from biowarfare in that they do not wish to risk a like-for-like retaliation. Significantly, such ideas of military value and deterrence *do* appear within this book in relation to biowarfare. Over the coming pages, the reader will see some actors framing their approach towards bioweapons based on calculations of military usefulness and the need to deter. This book will *also* show, however, that these arguments never win out entirely in respect to biowarfare policy. In taking this argument further, it is claimed that there is another explanation at work: the taboo. The book examines how actors have previously informed and justified their decisions not to employ bioweapons in order to demonstrate that decision-making is repeatedly driven by normative restrictions. The analysis shows that both the utility and deterrence arguments are inadequate as a sole explicative for biowarfare policies and provides a corrective to that approach in reference to the taboo. Critically, the book also accepts that the taboo is imperfect. A further key aim of this study is then to show how multiple factors including material issues, self-interest, and normative constraints can all play into a specific outcome. The book still proves, however, that the taboo *is* a key factor in respect to biowarfare employment and cannot be ignored.

Yet the bioweapons taboo is not simply a question of use. Limiting the analytic discussion of the taboo to use alone is reductionist. Previous studies of the taboos around other forms of WMD—such as nuclear arms—have been criticized for focusing exclusively on the issue of weapons' employment and so failing to consider the full impact of a taboo (Atkinson 2010; Eden 2010: 837; Walker 2010: 869). Taboos also pertain to matters concerning possession of such weapons and the power which armaments exert or are seen to exert. This book actively engages with these criticisms to look more widely at the ways in which bioweapons are understood as an expression of both political and military power, the motivations that drive their possession, and the meaning of bioweapons within international politics. Biowarfare is held up as a special type of threat as a consequence of the taboo. How does this normative prioritization then dictate the role that bioweapons play within global politics beyond their detonation? In answering this question, the book provides a deep understanding of the bioweapons taboo that moves the debate beyond considerations of use to analyse all aspects of what these armaments mean as a prohibited form of violence.

A new model of taboo

The book additionally answers a further core question: what is a taboo? Taboos are typically understood as something you are supposed not to do. Taboos are seen as socially constructed standards which actors should not transgress, whereby those standards express the rejection of specific moral, ethical, or social behaviours that have been deemed prohibitively unacceptable. To quote Hutton Webster (1942: 13, 17), taboos are 'prohibitions observed as customs'; 'an imperative thou-shalt-not in the presence of the danger apprehended'.

This book directly questions the concept's analytic understanding thus far and explicitly within the context of International Relations (IR). The study demonstrates that current definitions as to what a taboo 'is' are inadequate and fail to fully distinguish the concept from non-taboo forms of norm. The failure often facilitates a tendency—explicitly or implicitly expressed—to focus primarily on the above-mentioned characteristic of compliance, that is, the extent to which an actor adheres to a taboo. A compliance-centric approach is shown here to be analytically unrepresentative and that it sets unrealistic expectations for actor behaviour, especially where those expectations are then employed as the basis of quantifiable conceptual comprehension and study. Compliance *does* characterize understandings of taboo. Yet normative adherence must be contextualized to a much greater extent. We live in a world of social prohibitions, but not all are taboo. What distinguishes cannibalism (a taboo) from, say, poor table manners? There is a fundamental difference between eating a person and not using a napkin while you do so—and it is not simply the strength of normative observance. Consequently, while the concept of taboo is recognized as a key aspect of IR,[3] this study identifies a significant gap in our understanding of what it is that we are talking about.

The book responds to this critical need for a more comprehensive understanding of taboo by constructing an original definitional model.[4] The model establishes a set of three necessary and interrelated criteria which identify a taboo as a discrete classification of prohibition characterized by: *disgust*, *stigmatization*, and *fetishization*. A taboo is a prohibitive norm that formulates around a subject (object/person/act) which disgusts, whereby the response of disgust is manifest within the subject's social construction and rejection as a stigma, and the inherent rejection is expressed through, as well as further reinforced by an excessive value associated with, fetishization. If a prohibitive norm fulfils the terms of this definition it can be considered a taboo.

The new model engenders conceptual and methodological implications concerning the study of taboos within IR. In particular, an explicit focus on wider characteristics means that the original approach can hypothetically tolerate (at least some) examples of normative violation or considerations of violation, that is, whereby an actor may think or voice the

[3] The concept is applied in relation to a wide range of issues within IR including: killing civilians (Carpenter and Montgomery 2020); terrorism (Sluka 2008), including negotiating with terrorists (Toros 2008); torture (Frankenberg 2008); female representation within the military (Vermeij 2020); and prohibitions on certain types of weapon (Dolan 2013).

[4] The argument is based on a previous publication: Michelle Bentley (2022) 'A New Model of "Taboo": Disgust, Stigmatization, and Fetishization', *International Studies Review* 24(3), viac028: 1–25.

possibility of transgression but does not follow through on that potentiality. A compliance-centric method would logically dismiss all contravention as the non-existence of a taboo. This study, however, does not consider a lack of adherence to be necessarily problematic in respect of establishing that a taboo exists. In making this statement, the model innovatively opens up the analysis of taboos to consider behaviours and outcomes that are not now limited by a binary between pure obedience and pure compliance. Examples would include more conditional and secondary issues such as concession and compromise politics. Critically, this book still accepts that evidence of extensive and repeated violation may question the existence, strength, and scope of influence of a taboo—what Nina Tannenwald (2007: 369) terms being normatively 'robust'. Yet it is maintained that less-than-perfect observance does not *automatically* negate a taboo. Moreover, the book also demonstrates that violation can actually reinforce normative commitment.

The new model offers a more effective understanding of an extremely important concept within IR. Taboos demarcate an especially prominent type of norm and they exert significant influence in respect of international behaviour, not least where taboos are fetishized as such. Consequently, a taboo is not the same as a 'standard' norm and being able to comprehend its distinct power, presence, and function within international political behaviour is critical. Analysis requires an understanding of the concept which accurately portrays not only what a taboo is, but how a taboo performs—where the emotive response of disgust, the constructed stigmatization of the subject, and the fetishization of that subject as a priority concern all shape what this specific type of norm 'does'. Failing to adopt such a model risks leading to a paucity of debate on, what is, a key normative pressure within the international community.

An overlooked taboo

Extensive studies have already been published on the taboos relating to other forms of WMD. Analyses notably include Tannenwald's (2007) work on nuclear weapons and Richard Price's (1997) discussion of chemical warfare. Yet there has been no comparable examination of the bioweapons taboo to date.[5] While the presence of a taboo is recognized, there is no

[5] Philip McCauley and Roger Payne's (2010) article—'The Illogic of the Biological Weapons Taboo'—deserves an explicit mention here. The authors engage briefly with the history of biowarfare and the problematic nature of prohibition within the context of recognizing that a taboo exists. Yet the study does not discuss in depth what a taboo is or how this applies to biowarfare.

comprehensive assessment. Multiple reasons explain this gap. Repulsion can make biowarfare an unattractive and, therefore, overlooked topic. There also exists a widely acknowledged lack of data. State biowarfare programmes in particular are highly secretive. Consequently, 'it is difficult to answer even basic questions about the extent and character of BW [bioweapons] proliferation' (Carus 2017b: 129; see also Balmer 2002a: 120; Koblentz 2009: 4, 2013: 502; Barras and Greub 2014: 497; Parachini 2018: 91). Significantly, the clandestine nature of bioweapons policy and development is connected to the taboo. Concealment could be interpreted as a commitment to national security and maintaining military secrets. Yet the situation has instead been explained as a case wherein actors wish to avoid the social and moral opprobrium attached to these weapons by staying quiet (Conway 1972: 4; Hay 1999: 199; Carus 2017b: 136). Actors' silence may comprise not only a deceit to others, but also perhaps to themselves.

A further barrier to understanding the taboo relates to an overstated conflation of bioweapons with chemical armaments. To the limited extent that the bioweapons taboo has been acknowledged, the taboo is frequently equated with a similar normative prohibition against chemical warfare (e.g. Krickus 1965). Catherine Jefferson (2009) talks of the 'poison taboo' which expresses our normative rejection of *both* the biological and chemical means of aggression (see also Cole 1998). The conflation is not entirely unfounded. Critical parallels exist between the two types of weaponry. Both biological and chemical agents pollute the body and the fear of this contamination plays into these weapons' respective social construction as taboo. Moreover, it is only since the emergence of germ theory in the late 1800s that biological and chemical processes have been separated. Prior to this scientific development, political and legal conceptualizations of the threats often addressed biological and chemical violence together under the umbrella term of 'poison' in that the two types of weaponry were effectively considered the same. The connection, however, is now overplayed.

Two key disparities exist between biological and chemical violence. First, bioweapons are 'quite distinct' from their chemical counterparts (Spiers 2010: 12; see also Koblentz 2009: 5; Horowitz and Narang 2014: 528). Whereas bioweapons comprise living organisms that can be contagious, chemical weapons are non-living agents that cannot replicate (Kelle 1997: 138). Whereas bioweapons are slow acting (symptoms may take weeks to show and contagion can take time to spread), chemical violence is relatively quick (F. Smith 2014: 7). Whereas bioweapons are intrinsic to nature, chemicals are more suited to a laboratory (Kellman 2007: 12). To the extent that bioweapons are deemed more physically destructive compared to chemical arms, then the former are also considered more abhorrent and immoral (Hansen 1990: 55; Cole 1997: 5). It is argued on this basis that these two categories of armament must be treated independently (Koblentz 2009: 5; F. Smith 2014). Furthermore, each category of weapon has become associated with very separate arms control regimes: notably, the BWC and the 1993 Chemical Weapons Convention (CWC). Physically and socially, therefore, biological and chemical arms are not one and the same—and cannot be assumed as such in respect of taboos.

Second, a joint focus with chemical weapons skews analysis by effectively drowning out bioweapons from analysis. Chemical armaments have been more readily employed and data on them is less clandestine. Consequently, events and debates pertaining to chemical violence have received substantially more airtime within academic and policy discussions compared to biowarfare. Hence bioweapons are sidelined. Biowarfare is understood instead through a lens built around chemical issues that may bear no relation to biological aggression beyond indirect extrapolation. The predominance of the chemical threat means that the bioweapons taboo becomes both lost and misinterpreted in respect to the former. In contrast, this book treats the bioweapons taboo as an idiosyncratic issue within its own right. The analysis puts those overlooked issues of biowarfare in the spotlight and removes the problematic focus on chemical arms. Significantly, the book does not ignore chemical arms. The study examines key events and issues concerning chemical violence where these examples also impact upon the bioweapons taboo. The two forms of violence are connected and the impact that activity on one can have on the taboo of the other should not be denied. These chemical examples are included, however, only when there is a direct link to biowarfare. The book maintains that it is wholly insufficient to expect to find an answer to the bioweapons taboo through existing studies on chemical arms or investigations wherein those arms are equated to the biological. There needs to be a separate study of the bioweapons taboo on its own terms. This statement comprises a core rationale and an original contribution of the book.

A history of the taboo

The book is based on an historical and genealogical analysis. The study focuses primarily on a specific period of normative development between the Geneva Protocol and the BWC. These two landmark international agreements can be considered almost as the book ends of the political establishment and codification of the taboo. Critically, the selected time period is not a statement that the taboo failed to exist before Geneva or that it was perfected as soon as the ink had dried on the Convention. Nor are these agreements considered directly comparable. Nicholas Sims (2001: 152) points out that the Protocol is a 'law of war' and not a disarmament treaty like the BWC, whereby the former is significantly 'narrower' in scope. Nor are these agreements considered perfect and this study will highlight the flaws relating to them. The agreements are chosen, however, in that they (1) represent the explicit codification of the bioweapons taboo within international law as well as its customary status as such and (2) comprise bans on bioweapons that have truly cemented the taboo as an intrinsic aspect of international policymaking. Both agreements assist in explaining the taboo's development, meaning, and current normative robustness.

The book is structured around five historical case studies. The cases are not independent; the examples are interrelated, heavily contextualized, and provide an overarching historical narrative to the time period examined. Cases have been selected whereby they cover the following: core milestones in respect to the genealogy and evolution of the taboo; key examples of the threatened use, non-use, and rejection of bioweapons; and critical themes pertaining to the taboo, for example codification. Replicating Tannenwald and Price, the cases focus on the US. America provides an extremely apposite point of analysis in that the US has overseen the development of a major biowarfare programme, officially renounced bioweapons, and been repeatedly instrumental in developing international biowarfare prohibitions. Admittedly, this approach precludes a detailed examination of other examples relevant to the taboo; for example, the bioweapon programmes of Japan during World War II and the Soviet *Biopreparat*. Concentrating on the US also introduces a Western-centric bias, which is acknowledged here as a limitation on the study. The US case, however, still comprises a highly valuable example by which to examine the taboo. The US example is also generalizable to other actors and cases, as Tannenwald (2007: 23) argues of her own study on nuclear arms.

Data are drawn from extensive archival research based on the documentary research method, qualitative discursive analysis, and process-tracing through both primary and secondary sources. The concerns about the lack of data on biowarfare mentioned above still stand in this respect. The book, however, provides at least a partial corrective to this situation by offering an in-depth analysis of open archival records. Documents were primarily identified through official government sources (e.g. *Compilation of Presidential Documents*, *Foreign Relations of the United States*, *Department of State Bulletin*,

Documents on Disarmament, the CIA archives, as well as transcripts of Congressional hearings) and other US archives (National Archives, National Security Archive, and presidential libraries). The analysis also draws upon other forms of primary data, including autobiographies and media sources (especially the *New York Times* archive).

Structure and case studies

Chapter 1 outlines the new model of taboo proposed within this book in depth. The analysis explains the meaning and relevance of disgust, stigmatization, and fetishization to demonstrate how these factors inform a taboo. The chapter applies the model directly to the issue of biowarfare to show precisely why actors find these armaments to be taboo. Chapter 2 expands the conceptual analysis and establishes a methodology for identifying and measuring a taboo. How do we locate the presence and influence of a taboo? How do the specific characteristics of a taboo shape what we look for in respect of that taboo's identification? What types of evidence indicate that an actor adheres to a taboo? How does the rejection of a compliance-centric approach alter current methods for analysing taboos? In answering these questions, the book presents a critical analysis of Tannenwald's research in particular that contextualizes her claims within the context of the new model.

Chapter 3 studies the earlier history and codification of the taboo culminating in the 1925 Protocol. The chapter charts the evolution of the taboo's codification from the *Laws of Manu* c. 200 BCE, through various war treaties, the 1899 and 1907 Hague Conferences, to Geneva. The analysis depicts the growing formalization of the taboo not only as a socially established revulsion but also as a customary norm manifest within international legal provision. The chapter further documents the conceptualization of biowarfare alongside chemical warfare within the more general term of poison. As already highlighted above, bioweapons were initially addressed in tandem with chemical weapons. Yet the Protocol was instrumental in acknowledging biowarfare as a taboo issue within its own right and as distinct from the threat of chemical violence. The chapter explains this key shift in understanding.

Chapter 4 analyses the US bioweapon programme initiated during World War II. The existence of a major programme ostensibly disputes the taboo's relevance in that actors were willing to develop and possess biological armaments. The chapter demonstrates, however, that decision-making was significantly more partial and more deferent towards the taboo than might appear on the surface. The commitment to bioweapons was constructed only as the means of last resort, normatively mitigated to defence purposes only (whereby the offensive was rejected as morally wrong), and pursued in direct opposition to wider public and world opinion. Furthermore, the taboo undermined the pro-biowarfare standpoint to ensure both the decisive limitation of that position and a failure to incorporate bioweapons within military strategy. The taboo also shaped the terms of the debate in respect to rhetoric.

Chapter 5 considers the allegations of US bioweapons use during the Korean War. The study explores the influence of the taboo at this time with particular reference to the strengthening international perceptions of the prohibition at the United Nations (UN), how perceptions of violation reinforce the taboo, and expression of the taboo within strategic narratives. The chapter also assesses a further potential challenge to the taboo's relevance: the clandestine introduction of a first-use biowarfare policy in 1956. The policy could be interpreted to undermine the taboo. This first-use commitment, however, is contextualized within the book to show that: the policy was never exploited; there was never any threat that it would, as a consequence of the taboo; the ultimate powers-that-be continued to express their commitment to the taboo as a priority; and even the hint of offensive biowarfare activity was met with extensive political and public opposition outside the secretive military context in which the strategy originated.

Chapter 6 examines President Richard Nixon's 1969 renunciation of bioweapons. The case comprises a crucial example whereby it represents an explicit and total rejection of biowarfare. Renunciation is typically attributed to self-interested calculations of military utility. In contrast to this accepted thinking, the chapter offers an original and alternative interpretation of events based on the taboo. The analysis identifies a range of causative factors pertaining to the taboo: scandal relating to biowarfare, for example Vietnam; public opinion, which was galvanized by (what is termed within this book) 'taboo advocates'; and the continued international stigmatization of biological aggression. The chapter also considers the dynamics between military and wider public perceptions of biowarfare to ascertain how transparency relates to the taboo. It is demonstrated that the less secret biowarfare policy becomes, the more the taboo takes hold.

Chapter 7 analyses the BWC. The chapter contests existing claims that American acceptance of the BWC was purely self-interested in motivation. A close analysis of the archival evidence demonstrates that the taboo comprised a major factor in the US's negotiation and ratification of the agreement. Moreover, the data reveal that a key rationale of the BWC was to cement the taboo as an intrinsic norm of international policymaking. The chapter contextualizes this argument in respect of criticisms that the BWC is not effective nor some great expression of normative feeling. Indeed, the book does not argue that the BWC represents *full* robustness of the taboo. It is maintained, however, that the agreement still constitutes the current mainstay of the taboo's genealogy. The BWC comprises the point at which the norm was truly enshrined as a core dynamic of international politics. The BWC embodies not only a customary history within which the taboo had shaped security decisions for centuries before; the document also represents an on-going commitment to the taboo—which we still see in the non-use and rejection of bioweapons today.

1
A Bioweapons Taboo

> What is taboo is revolting, untouchable, filthy, unmentionable, dangerous, disturbing, thrilling—but above all powerful.
>
> **(Burridge 2004: 199)**

Derived from the Tongan *tapu/tabu*, the origins of 'taboo' within the English language are attributed to the explorer Captain James Cook (Eden 2010: 831). Cook (1967: 176) described taboo as 'a word of an extensive signification'. He initially associated the concept with customs concerning the rejection of food, whereby certain foods and types of food preparation were considered impermissible. Cook had documented such behaviours on previous voyages, although he did not then refer to these as taboo. He recounted an incident in Tahiti on his first world voyage between 1768 and 1771 wherein females were not permitted to eat with the men in a community. Instead the women 'would Often goe 5 or 6 together into the Servants apartments, and there eat very heartily of whatever they could find' (J. Cook 2014: 91). When asked why, the females replied, 'that they did it because it was right, and Express'd much dislike at the Custom of Men and Women Eating together of the same Victuals'. Food also relates to Cook's first explicit reference to a taboo. Cook described inviting several local dignitaries for dinner on his ship in a diary entry for 15 June 1777. Cook (1967: 129) reported: 'When dinner came on table not one of my guests would sit down or eat a bit of any thing that was there. Every one [item of food] was *Tabu*.' Yet Cook also discussed the relevance of taboos to the sacred. For example, the word was used within the context of human sacrifice and, in a further revelation, 'they tell us that if the King [the sacred party] should happen to go into a house belonging to a subject, that house would be Taboo and never more be inhabited by the owner' (Cook (1967: 176). Captain James King (1793: 333)—who recorded the rest of Cook's final voyage after the explorer's death on it—likewise observed that a taboo was 'used to express any thing sacred, or eminent, or devoted'.

The Biological Weapons Taboo. Michelle Bentley, Oxford University Press. © Michelle Bentley (2023).
DOI: 10.1093/oso/9780198892151.003.0002

Consequently, the concept of taboo can be seen to comprise two key elements: the unclean and the sacred. Anthropologist F.C. French (1907: 206) summarizes these two interrelated classes of taboo, whereby there are: 'Things too sacred to touch, on the one hand, and things too impure to touch, on the other.' The distinction also underpins the understanding of taboo employed by Sigmund Freud in his seminal work, *Totem and Taboo*—an analysis of 'primitive' taboos and their connection to religious and sacred ritual. As Freud (2012: 13) expresses it, these two forms of taboo represent different 'directions' as to what is improper but that there constitutes no distinction between them where both comprise a taboo. James George Frazer (2013: 203) argues in his 1890 book, *The Golden Bough*, that these two strands of taboo are united by the danger that both pose to anyone who violates a prohibition. Danger is so engrained within the notion of taboo that Franz Steiner (1956: 20–1) asserts that 'taboo deals with the sociology of danger itself'. The implications of provoking such danger by coming into contact with what is taboo are considerable. Keith Allan and Kate Burridge (2006: 5) even talk of 'fatal taboos', whereby a violation can (or be thought to) lead to death in the absence of punishment or purification. Frazer (1875: 17) even comments that: 'Cases are on record in which persons who had unwittingly broken a taboo actually died of terror on discovering their fatal error.'

Taboo perceptions of danger may not be rational (Toy 1899; French 1907). The idea of danger expressed here is specifically a 'mystic' or 'mysterious power' (Marett 1914: 81; Burriss 1929: 144). Freud (2012: 4, 14) establishes that there is no clear rationality in respect of the motivations inherent to a taboo as a parallel to Kant's Categorical Imperative. A taboo is not constructed in terms of its own justification; it exists in the absence of any rational imperative for that existence (Feinberg 1988: 21). Perceptions of irrationality, however, must be examined alongside the work of Gertrud Koch (2003: 78), who comments: 'When considered as a technique, the taboo can be said to have a rational core: it is functional with the practical goal of damage prevention.' Taboos are performative as a type of social control in that they deter persons from engaging in behaviours that are themselves deemed dangerous. Yet a lack of rationality does not negate a taboo's existence or impact. As Frazer (2013: 203) argues in relation to the 'spiritual or ghostly, and therefore imaginary' danger associated with taboo: 'The danger, however, is no less real because it is imaginary; imagination acts upon man as really as does gravitation, and may kill him as certainly as a dose of prussic acid.'

There is a lack of consensus on what a taboo is within anthropological circles. Nor has the concept always escaped its own contextual and relativistic interpretation. Freud (2012: 13) argued that the true definition of a taboo was lost in its English translation and furthermore: 'Taboo is a Polynesian word, the translation of which provides difficulties for us because we no longer possess the idea which it connotes' (see also Boivin 1973: 102–3). Alfred Radcliffe-Brown (1939: 8) uses both 'tabu' and 'taboo' in an attempt to distinguish between the Polynesian and English versions. Yet even he acknowledges that his approach 'is not entirely satisfactory'. Steiner (1956: 50–1) claimed that the situation is complicated even further by the way in which the Polynesian taboo was effectively re-problematized during the Victorian era. Steiner comments that the word became associated with that specific 'taboo-ridden' society whereby the likes of Frazer 'grew up among people who preferred, in certain circumstances, to say "unmentionables" instead of "trousers"'.

Previous anthropological interpretations can also be accused of Orientalism in the vein of Edward Said (2003). In further explaining why the concept of taboo is now so far beyond our comprehension, Freud (2012: 15) stated that this is because 'no understanding of it could be possible without entering into the study of the belief in spirits and demons which is so characteristic of these *low grades of culture* [emphasis added]'. Crucially then, Freud's differentiation is not simply a case of cultural or social distinction between separate groups of people, but an assessment based on the denigration of the other. The other is portrayed as someone of simple mind. Webster (1942: 8) similarly says that primitive taboos are the product of 'a lively imagination and of an abysmal ignorance'. Such reasoning constructs an unflattering picture of a savage whose 'brain is not developed physiologically for him to make proper associations and come to the right conclusions' (Burriss 1929: 144). Some may feel that this portrayal still applies to many political actors today. Yet assuming that a taboo comprises a primitive construction by intellectually inferior groups undermines the very concept by writing it off as something which civilized people do not engage with.

Stripping back a taboo

Margaret Mead (1937: 502) contends that these conceptual problems and dissonance can be overcome by extracting a fundamental notion of taboo from its anthropological context. The concept 'must be stripped of those accidents of interpretation'. In doing so, Mead establishes that a generalizable conceptualization of taboo can exist separate to, and removable from, the narrow and provocative associations of anthropology. The idea of a cross-disciplinary model of taboo is extremely valuable where it can then be related to other areas of study—notably here, IR. It is unviable to simply take the anthropological understanding of taboo and apply it 'as is' to IR. This approach would exceed the disciplinary relevance of the concept and expect that conceptualization to measure behaviour in ways that it was never designed to. Mead, however, opens up an opportunity to consider taboo within alternative contexts. Critically, this statement is not to suggest that taboo has an essentialist meaning. It is instead argued that aspects of the anthropological understanding can be drawn upon to construct a workable model of taboo applicable beyond its original disciplinary frame.

Current IR approaches are inadequate in terms of realizing such a conception of taboo. Specifically, the discipline has yet to fully articulate how a taboo is distinguished from other norms. The social construction and integration of taboos has seen the concept categorized as a type of norm. To define norms: 'Norms are regarded as standards of behaviour, defined in terms of rights and obligations. In this sense norms are general prescriptions of behaviour that regulate intentions and effects' (Bjorkdahl 2002: 13). Normative behaviour can be understood with respect to Peter Katzenstein's (1996: 5) accepted conceptualization of 'the proper behaviour of actors within a given identity'. Martha Finnemore and Kathryn Sikkink (1998: 891) similarly adopt the idea of 'proper' conduct as the basis for normative construction. Simply put, norms constitute the socialized 'rules' that shape actor behaviour. IR analysis frequently distinguishes taboos from other norms in that the former (1) express a type of prohibition, also referred to as a 'prohibitive norm' (Alolaimy 2017: 43)[1] and (2) achieve a 'near-universal' level of compliance (Press, Sagan, and Valentino 2013: 190). Compliance is understood to be fundamentally automatic and instinctive. As Tannenwald (2007: 10, 11) describes it, a taboo comprises 'a particularly forceful kind of normative prohibition' that is 'associated with such qualities as absoluteness, unthinkingness, and taken-for-grantedness'. Adherence happens unconsciously and—says T.V. Paul (2010: 857)—that an actor would not even *consider*, let alone perform, a violation if a taboo is truly a taboo.

Consequently, compliance is often at the heart of defining taboo within IR as a primary analytic standard. For example, Harald Mueller (2021: 1082, 1083) considers a taboo to comprise a norm which specifically 'prohibits something in absolute terms' and whereby taboos are powerful explicitly because of 'their absolute validity'. Mueller (2021: 1084) understands a taboo as demonstrating 'virtually irresistible strength', and that this trait gives the

[1] This book uses the term 'prohibitive norm' to describe a norm that expresses a prohibition, but which may or may not be associated with a taboo. A prohibitive norm can refer to a taboo but can also refer to a norm that proposes a prohibition, but which does not portray the specific characteristics of a taboo, e.g. the poor table manners mentioned in the introduction.

taboo 'distinctive meaning' from any non-taboo norm. He concludes by this reasoning that if a taboo is not adhered to then it should not be considered as such. Mueller (2021: 1085) specifically challenges the nuclear taboo—the claim that these armaments have not been employed, at least in respect of a first strike option, because they are normatively prohibited (e.g. Tannenwald 2007). Mueller questions whether the nuclear taboo exists on the basis that such a restriction 'is possible only in a country with zero nuclear weapons and no membership in a nuclear alliance, and universally only in a world without nuclear weapons'. Mueller uses the total rejection of nuclear armaments (i.e. full compliance) as the standard by which a taboo can be identified. The same approach is evident within the work of Rebecca Davis Gibbons and Keir Lieber (2019: 32). The authors reject the reference to the nuclear 'taboo' and prefer the label 'norm of nuclear non-use' due to that norm's actual or desired violation—whereas 'real' taboos would encompass more definitive prohibitions that are only rarely subject to transgression, for example cannibalism. Once again, compliance drives the ultimate decision as to ascertaining the presence of a taboo.

Similarly, Daryl Press, Scott Sagan, and Benjamin Valentino (2013) contend in their landmark study that nuclear arms are not taboo due to evidence of violation. They instead call non-use a 'tradition' (see also Paul 2009, 2010). The authors conclude that people (in this case, the American public) claim to be willing to employ nuclear armaments under certain hypothetical circumstances, for example in response to a terrorist attack. Consequently, the participants' 'weak aversion' demonstrates that non-use 'has none of the bright-line, absolute, or unthinkable qualities of a taboo' (Press, Sagan, and Valentino 2013: 188, 189). Non-use is not the result of overpowering normative adherence drawing on 'moral reservations, concern for America's reputation or a sense that the weapons are inherently uncivilized' (i.e. taboo). Non-use is determined by actors who do not wish to break social expectations that these armaments should not be employed in that this behaviour comprises precedent. Avoidance of nuclear weapons is not a taboo; it is habit. In making this statement, the authors' identification of a taboo is highly dependent on compliance as an analytic standard. While they negate the presence of the taboo on the basis that other motivations apply, the negation is still premised on a lack of compliance with the taboo as a consequence of those motivations. A taboo must be 'bright-line, absolute, or unthinkable' to qualify. Admittedly, there exists an element of concession in that the authors attest that taboos are said to demand only 'near-universal' compliance, that is, not complete (Press, Sagan, and Valentino 2013: 190). Yet they are still seeking to identify a taboo as a norm that demands adherence. Indeed, Sagan and

Valentino (2021: 1093) elsewhere accept Mueller's argument that there are 'no circumstances permitting an exception [i.e. violation] to the prohibition [if the taboo is to stand]'.

There are two key issues with a compliance-centric approach. First, the focus on compliance raises the question: what taboo is total? Taboos *are* violated. People *do* have sex with family members, have sex with children, and eat other people. No one doubts, however, that incest, paedophilia, and cannibalism are taboo. The stark reality is that not all taboos are respected and enforced to the same or total degree. Nor should they, as Lawrence Freedman (2013: 97) highlights in his discussion of the civil rights movement as a valuable challenge to racist taboos.[2] Similarly, Jessica Auchter (2017: 232) discusses how breaking the taboo around seeing and publicizing images of a corpse can ultimately help draw public attention to critical political issues. Taking this point a step further, it is conceptually impractical to measure 'near-universal' behaviour. How many people have to be scoffed down for dinner before cannibalism stops being taboo? It is also premature to dismiss cases wherein an actor considers transgression but does not act upon that impulse, per Paul. The assumption necessarily ignores the possibility that a taboo still ultimately restrains. Actors may think of doing 'bad' things, but they may not do them to the extent they are, or become, reconciled to the idea that the act in question is undoable if not entirely unthinkable. The concept of the 'unthinkable' comprises a handy means of expressing the excessive repulsion associated with a taboo. It is unhelpful, however, to take this idea in its most literal form when identifying a taboo itself.

Second, other scholars working on IR taboos do recognize that there is 'more' to a taboo than its own compliance. The approach frequently draws upon anthropological understandings of taboo as a type of danger. For example, Tannenwald (2007: 10–11) speaks of the 'sociology of danger' to express that there is something fundamental to a taboo which goes beyond normative adherence. In particular, Tannenwald deems the implications of violation to be more excessive and transformative than those attached to the violation of a non-taboo norm. She further argues that the social construction of a taboo can be more important than compliance. Tannenwald (2007: 10) says of nuclear aggression: 'The [nuclear] taboo is not the behavior (of non-use) itself but *the normative belief about the behavior*' [emphasis in original]. She explains away non-compliance, as well as disparities in normative robustness between different taboos (where she accepts, for example, that the

[2] Walter Gantz (1978) says: 'Taboos can protect but can also destroy. They can help maintain a society's stability or can break down its social fabric. When backward-orientated, they retard social change and development as well as individual growth and psychological comfort'.

nuclear taboo is 'more fragile' than others such as incest (Tannenwald 2007: 16), by focusing on the *perception* of perfect compliance as opposed to actual perfection. If people (are socially constructed to) believe a taboo to exist then it does so, even in the face of violation.

Yet there are two issues with even this more expansive approach. First, Tannenwald's understanding still risks (at least unintentionally) a focus on compliance whereby she discusses the taboo primarily in terms of strength, per the 'particularly forceful' reference above. A taboo is defined according to the potency of the danger involved. The reference to strength implicitly reduces the concept to the degree to which danger is felt and responded to. Consequently, the analysis then again translates into an issue of compliance by default. To the extent that a taboo is recognized as a 'bright line' norm, this reference is instantly associated with the absolute and unthinkable—both measures of adherence. Identifying the taboo in terms of its extreme voracity, and whereby taboos are then judged with transgression at the forefront (see also Gizweski 1996: 399–400), indicates compliance because the taboo is assessed primarily according to whether it has the capacity to constrain.

Furthermore, Tannenwald limits her conception of the nuclear taboo to first-strike options only. Given that states do possess nuclear arms, and that possession is actively connected to notions of deterrence (whereby a second-strike capacity is effectively sanctioned as a permissible, if conditional, strategy), Tannenwald restricts the taboo to certain nuclear behaviours. Her restriction can be attributed to an inherently compliance-centric assumption. Tannenwald still ultimately treats normative adherence as the primary basis as to whether a taboo exists—whereby she reasons that, if states choose to retain weapons, then the taboo must be conditional.[3] Crucially, the observation here is not to overstate the compliance argument in respect to Tannenwald. Her approach includes an analysis of wider issues pertaining to the taboo beyond compliance, such as the perceived morality of nuclear weapons. Not to clarify this point would comprise a misrepresentation of Tannenwald's research and the important contribution that she makes to the debate. It must also be fully acknowledged that Tannenwald is open to the toleration of violation in other aspects of her analysis. Tannenwald (2007: 16) says of strong taboos such as incest that: 'Numerous violations … can occur, and they [the taboos] remain fairly resilient.' Interestingly, however, she does not apply this caveat to the more 'fragile' nuclear taboo.

[3] It should be noted that Tannenwald (1999) did not limit the taboo to first-strike options in her earlier work on the nuclear taboo.

The identified bias towards compliance is perhaps unsurprising given that normative adherence is still considered by many to comprise the core analytic standard. For example, Alexander Montgomery and Charli Carpenter (2021: 1089) have started to question absolute notions of taboo based on compliance. In looking at the nuclear taboo's violation they state: 'These scenarios demonstrate logics that can enable exceptions to the nuclear taboo without making it less taboo-like.' Yet this claim has come under severe fire from Sagan and Valentino (2021: 1093), who state that the authors 'are simply misusing the term "taboo" here', whereby they maintain that non-compliance is 'evidence against there being no taboo at all'. Breaking down the standard of compliance is controversial.

Second, even in cases where scholars are open to a wider conceptualization of taboo as an intrinsic danger, this approach has yet to fully articulate what this danger entails. Danger may be a reason for compliance: that is to say, it is because actors perceive a taboo as dangerous that they adhere to it. Yet what comprises that danger? To put the matter another way: if a taboo can be violated and still be 'taboo-like'—per Montgomery and Carpenter—what *makes* it 'taboo-like'? What sustains a taboo in the face of violation as Tannenwald suggests above? Tannenwald fails to explain what makes a taboo sufficiently compelling that it can withstand violation and still comprise a taboo. The issue feeds into a further, related, issue as to what distinguishes a taboo from other forms of non-taboo prohibitive norm. Actors are subject to numerous prohibitions, but they are not all necessarily taboo. We do not walk into the road when the crossing light is red because we have been taught expectations of road safety. We do not take the last biscuit out of propriety. We do not tell someone their jumper is horrendously ugly in an effort to be polite. Yet we would not consider these behaviours to be the consequence of a taboo. What makes a taboo stand out? On what grounds do we know when we are dealing with—indeed, that we are influenced by—a taboo explicitly?

Consequently, this study has identified a critical need for a more detailed understanding of taboo which (1) contextualizes the definitional reference to compliance and (2) explains how the inherent danger of a taboo is manifest. In response, a new model of taboo is put forward based on three necessary and interrelated criteria: *disgust*, *stigmatization*, and *fetishization*. Taboo comprises a separate category of prohibitive norm that is distinctly: visceral; caused by and generating disgust; stigmatized as a means of social control; enforced by sensations of shame; fetishized according to a construct of excessive value; and attracting extreme and emotive responses.

Disgust

Rejecting assertions that emotions are either irrelevant or so irrational as to be analytically immeasurable, there has been a so-called 'emotional turn' within IR. The approach argues that emotions comprise a valid and significant element of international politics (Fattah and Fierke 2009: 69). It is now widely recognized that both emotions and emotional belief determine actor behaviour and explain how actor motivations are shaped by their emotional experience, including where this relates to moral judgement (e.g. Hymans 2006; Bleiker and Hutchinson 2008; Sasley 2010; Erskine and Lebow 2012; Jeffrey 2014; Mercer 2014; Reus-Smit 2014; Ross 2014; Hall and Ross 2015; Reinke de Buitrago 2016). As Neta Crawford (2013: 121) asserts: 'Emotions are everywhere in world politics and we ignore them at our peril.' Emotion applies even to the extent that IR theory is constructed around the state as opposed to an individual. Brent Steele (2008: 16) demonstrates that emotion pervades all paradigmatic strands of IR theory and insists that '[t]here is no such thing as the "cold, calculating" nation-state'. Emotions do comprise an intrinsically complex unit of analysis that is considered potentially difficult to study—not least the way in which emotions are 'undefined' in scope (Scheff and Retzinger 2002: 3; see also Elster 1999: 241) and where they are seen to come into conflict with competing notions of affect.[4] The role of emotion in international behaviour and identity construction, however, must still be addressed.

Within this theoretical vein, taboo is related here to the emotion of *disgust*. A taboo concerns a subject that explicitly elicits disgust. Disgust has been defined in a plurality of ways (Strohminger 2014: 481). Yet the emotion is fundamentally conceived of as a strong negative response to the unpleasant. Disgust constitutes a repulsed rejection of what is considered abhorrent. It is the nastiness that we cannot wait to get away from and which causes a retch at the back of our throats. Disgust is the putrefaction, the filth, and the sickening horror. As Winfried Menninghaus (2003: 1) expresses it: 'Every book about disgust is not least a book about the rotting corpse.' More explicitly, disgust comprises a rejection of that which involves the contamination of the body—what Daniel Kelly (2011: 19) terms 'contamination sensitivity'. Sensitivity, moreover, is not merely provoked by the subject that is disgusting as a source of contamination. A response is further elicited through the

[4] Eric Shouse (2005) distinguishes between the two whereby he identifies affect as 'a non-conscious experience of intensity ... a moment of unformed and unstructured potential' and emotion as 'the projection/display of a feeling'.

idea that this contamination can be passed on, where this relates back to the work of Frazer (2013: 204). For example, people refuse to consume a drink which has previously come into contact with a (sterilized) cockroach, even where they do not encounter the insect directly (Rozin, Millman, and Nemeroff 1986). Sensitivity also engenders the rejection of subjects similar to that which we find disgusting.[5] The reference to contamination is extremely important whereby it establishes why disgust constitutes more than a simple threat to the body and to life. We fear—but are not disgusted by—the knife thrust at our throat, or the gun to our chest. Disgust concerns not merely the protection of the body, therefore, but protection from harm explicitly as a contaminant.

Disgust is argued to be simultaneously biological *and* socially constructed. In terms of the former, disgust-as-rejection is comprehended as an evolutionary and instinctive reaction to safeguard the body (Gorman 2012). As Jesse Bering (2013: 24) points out: 'We do not have to learn how to vomit, for instance, after wolfing down a burger infected with *E. coli*.' Admittedly, the exact locus of biologically driven rejection is contested. All of the proposed explanations, however, still relate to the same general conceptualization of contamination sensitivity and disgust as a biological barrier to contaminants. Some analysts attribute this response to an oral rejection related to food (Rozin and Fallon 1987; Rozin, Haidt, and McCauley 2008). This interpretation draws on Charles Darwin's (2007: 269) assertion that disgust 'in its simplest sense, means something offensive to the taste'.[6] With echoes of Cook's resistant dinner guests, Paul Rozin (1997) provides an example: 'The offensiveness of eating worms is one of the most powerful things you can imagine.'[7] Other analysts, however, conceptualize biological rejection as a form of protection against disease (Curtis, Aunger, and Rabie 2004; Curtis 2007, 2011; Oaten, Stevenson, and Case 2009; Davey 2011: 3453; Schaller and Park 2011).

Disgust has also been defined as a fear of our animal nature (Rozin, Haidt, and McCauley 2008: 760, 761–2); more explicitly, the 'problematic relationship with our own animality' (Nussbaum 2006: 89). The claim builds on Andras Angyal's (1941) argument that disgust is primarily inspired by waste and animal products. Reminders of our animality are (1) non-human and (2) act as an uncomfortable indicator of our own mortality, vulnerability, and

[5] An extremely common example used in the literature is imitation dog faeces made of chocolate fudge.

[6] Other analysts base disgust on smell, e.g. Aurel Kolnai (2004a: 48, 50). Kolnai accepts, however, that there is a link between smell and taste.

[7] As the old joke goes, what's worse than biting into an apple and finding a worm? Biting into an apple and finding half a worm.

base nature. Opposed to the idea of being reduced to the status of animal, we reject anything which connects us to the non-human as a type of contamination. It is no coincidence that Edmund Leach (1964) has shown that references to animals are often employed as social insults, for example 'you dog', 'you beast'. Animality further reflects Julia Kristeva's (1982: 13) work on abjection. Kristeva discusses the experience of horror as an avoidance of that which is associated with 'the threatening world of animals or animalism'. Critically, the claim of animality is not unproblematic. Martha Nussbaum (2006: 89) notes that we also value the traits we share with the animal kingdom, such as strength and physical agility. Expanding upon Nussbaum's point, it can also be highlighted that we do not find all animals intrinsically disgusting. Fluffy kittens are extremely cute. Consequently, Nussbaum (2006: 92) appends a caveat: animality applies only to animal qualities and excretions that are *associated with death and decay*. Nussbaum (2006: 88) comments that we believe that 'if we absorb or are mingled with the decaying, we will ourselves be mortal and decaying'.[8]

Disgust and immorality

While some analysts continue to insist that disgust is so biological that it should not even be considered an emotion (Panksepp 2007), disgust is also claimed to be elicited as a social construction (as Bering goes on to argue; see also Kekes 1992: 436; W. Miller 1997).[9] Paul Rozin, Jonathan Haidt, and Clark McCauley (2008) speak of biologically driven rejection as 'core' disgust in the sense that this evolutionary manifestation has then translated into a much wider range of socialized disgust-related behaviours. These behaviours relate to highly diverse issues such as sex, magic, and death—and do not necessarily require a biological contaminant (e.g. food) to inspire revulsion. Kelly's (2011) 'co-opt thesis' follows the same line of thinking. Significantly, the socialized element explains a variation across cultures as to what is disgusting. For example, different foods are considered repellent within different

[8] Jefferson (2009: 69) argues that Nussbaum's concept of animality does not clearly engage with the idea of contamination. Nussbaum (2006: 93), however, repeatedly expresses her conceptualization in such terms, e.g. where elicitors of disgust are 'seen as contaminants because they are seen as reminders of our mortality and animal vulnerability'.

[9] Very young children do not tend to display disgust traits. Valerie Curtis and Adam Biran (2001: 27) suggest on this basis that the disgusted emotional response is developed later through socialization. The authors also caution, however, that this evidence does not necessarily mean that disgust is not innate in that children also have protection from parents. Consequently, children may have not yet developed disgust-related defences against contamination whereby they can rely on the caregiver for such protection at this stage.

cultures (Korsmeyer and Smith 2004: 2). While the Icelandic may enjoy eating hakarl—rotten aged shark that has been buried in a pit until the internal fluids have seeped out—it is not to everyone's taste.

Consequently, disgust not only relates to the physically unclean but also to the socially unclean—the uncivilized and socially deviant (Kekes 1992, 431; Horberg et al. 2009: 963). Norbert Elias (2000) argues in his famous 1939 book, *The Civilizing Process*, that the concept of disgust has developed in tandem with society's civilization; in particular, disgust has informed the social institutionalization of politeness and manners. Within a similar vein, Carolyn Korsmeyer and Barry Smith (2004: 23) describe disgust as not merely as a reaction to biological contamination but to 'moral decay and foulness of character'. While the extent to which it is specifically disgust that motivates a negative response to perceived social wrongs has been challenged (Nabi 2002), Rozin, Haidt, and McCauley (2008: 763) maintain the connection. The authors assert that—although this linkage may be difficult to quantify—people 'apparently *feel* some similarity in their emotional reactions to feces and to sleazy politicians [emphasis in original]' (see also Haidt et al. 1997).

Disgust then acts as a 'gatekeeper' and a protector to the cultural and socially constructed self (S. Miller 2004: 5; see also Nussbaum 2006: 78; Rozin, Haidt, and McCauley 2008: 771; Plakias 2013). Furthermore, the emotion is often considered essential not only to precluding a specific act of immoral behaviour but also in preventing the wider social disruption that such an act may inspire through contagion, per the cockroach (Devlin 1968; Kass 1997; W. Miller 1997). Martha Nussbaum (2006) counters this view to claim that disgust should not inform processes of law-making. She asserts that this emotion is too irrational to act as a reliable moral barometer. Despite Nussbaum's concern, however, disgust is still employed here to identify that which can be taboo by virtue of the fact that it *does* disgust—including where this relates explicitly to moral behaviour. The social perceptions associated with disgust-as-rejection can be extreme and highly emotive. Menninghaus (2003: 1) comments: 'Everything seems at risk in the experience of disgust. It is a state of alarm and emergency ... in which what is in question is, quite literally, whether "to be or not to be"'. Within this context, disgust is applied here to explain not only the rejection inherent to a taboo (whereby a taboo embodies the censure of a prohibited subject), but also the sheer intensity of feeling that characterizes the prohibition. Both of these functions do not apply to non-taboo norms, at least to the same decisive extent. Subsequently, these factors comprise a key point of conceptual differentiation and definition. Disgust sets a taboo apart.

The disgust of disease

Bioweapons are taboo in that they are potential mass killers. Classified as a WMD, the sheer devastation that these armaments could inflict has long been considered self-evident in respect of pandemics such as the fourteenth century Black Death, 1918 Spanish Flu (Bentley 2014: 10), and now also Covid-19. The extreme destruction associated with these events is effectively transposed onto perceptions of biowarfare. The bioweapons taboo, however, is not simply a case of the extent of destruction that these armaments can cause. The taboo also relates to the *type* of destruction involved.

The type of destruction associated with bioweapons pertains to the disgust of disease. As already highlighted, disgust is intimately connected to notions of bodily contamination. Consequently, disease comprises a perfect example. Disease exudes a form of contamination that has 'almost inevitably been associated with pain, suffering, and tragedy' (Westing 1985a: 631; 1989: 81). Disease additionally relates to the concept of animality. Our base physical vulnerability to sickness leaves our bodies like those of animals—prey to nature. Biological infection also evokes especially extreme sensations of repulsion that exceed those associated with other types of contaminant. One academic study measured disgust responses by showing participants images associated with revulsion. Participants elicited a stronger reaction to pictures explicitly related to disease than those with little or no connection (Curtis, Aunger, and Rabie 2004: S131). It is also not irrelevant that the core physical reactions of disgust—for example, the impulse to vomit—feel a lot like being unwell. The deliberate contamination of bioviolence is even more terrifying (Stern 2002: 104). Biowarfare concerns not simply the horror of disease, but a case whereby this abhorrent contamination is coercively forced upon the body as an act of calculated aggression.

Disease then elicits a psychological trauma, which is considered an intrinsic aspect of bioweapons' effect. Fear of disease can generate mass panic on a debilitating scale. Biological armaments are termed 'weapons of terror' on this basis (Wessely, Hyams, and Bartholomew 2001: 878). The nature of this anxiety is such that bioweapons would not even have to be employed to create a negative psychological effect. The dread of use alone is sufficient. As Mahatma Gandhi (2009: 70) famously expressed: 'Fear of disease killed more men than disease itself.' Per the anthropological definition of a taboo, the fear concerning disease is not necessarily rational. For example, Jessica Stern (2002) speaks of bioterrorism within the context of 'dreaded risks.' She discusses actors' response to the bioterrorist threat as disproportionate compared to the actual risk and attributes this excessive reaction to a deeper, visceral, and emotive fear of the disease involved. Irrationality is further reflected within the way that the social construction of certain diseases inspires more fear than others even though they may cause equal or lesser destruction (Enemark 2007: 131). Malaria is a major killer yet can generate less alarm than seemingly more exotic, but significantly rarer, diseases such as Ebola. Furthermore, it is unclear why actors are more concerned about sickness than alternative forms of death or injury. Disease *is* horrific. Getting blown up in a conventional explosion, however, is also hardly desirable. As Stephen Walt (2013) said of the greater fear surrounding chemical arms relative to forms of conventional aggression: 'Dead is dead—no matter how it is done.' Disgust, however, still drives actors to a perception of biowarfare as distinct; as a taboo.

Stigmatization

How does that which disgusts become manifest as a taboo? The answer lies in the work of Mary Douglas (1984); or rather, Nussbaum's critique of Douglas. Douglas interprets disgust as dirt. Dirt is understood here not as a specific issue of hygiene but as anomaly. Dirt is that which is out of place or defies 'normal' classification. The cockroach is not intrinsically disgusting until it has actually been swimming in your drink. Until the cockroach poses a contaminant threat, it is merely an insect. Douglas has attracted criticism for reductionism (W. Miller 1997: 44) to the extent that Nussbaum (2006: 91, 94) rejects this theory as a viable discussion of disgust generally. Nussbaum does, however, find Douglas valuable in discussing disgust vis-à-vis the idea and societal function of a taboo. Specifically, Douglas' model expresses the role of taboo in keeping 'dirt' in place and preventing any contamination from that which is taboo to the exposed person(s). Yet it is argued here that Douglas' model requires a further level of conceptual theorization to elucidate exactly how disgust (as the desire to keep matters in their rightful position) translates into a taboo: *stigmatization*. Disgust explains the causal element; stigmatization models the negative social response to this and distinguishes a taboo from simple disgust.

Originating in Graeco-Roman culture, stigma was a physical brand burned, cut, or tattooed into the skin in order to mark a person out in some unfavourable way; for example, as a slave or criminal (C. Jones 1987).[10] As in Hester Prynne's infamous red 'A' in *The Scarlet Letter* that signals her

[10] While the concept of stigmatization still draws upon its ancient meaning, it now focuses considerably less on a specific physical manifestation. Stigma can now relate to non-physical characteristics such as perceived weaknesses of 'individual character' and socially identified immorality (Goffman 1990: 14).

adultery (Hawthorne 1850), stigma comprises a visual and public communication of low social status. Seminal stigmatization scholar Erving Goffman (1990: 12) prefers to term this status as 'social identity'.[11] Significantly, the letter that Prynne is forced to adorn is not merely a symbolic representation of her socially impermissible act. The letter also facilitates her identification as someone to be isolated and condemned as an outcast and where her social alienation comprises a key aspect of her punishment. Stigma, therefore, not only relates to the stigmatizing feature. Stigma also concerns a subsequent societal exclusion that impacts upon the stigmatized person's ability to participate within a given society (Link and Phelan 2001). Goffman (1990: 15) describes the situation as a case wherein 'an individual who might have been received easily into ordinary social intercourse possesses a trait that can obtrude itself upon attention and turn those of us he meets away from him'.

To phrase this issue of rejection another way: stigmatization relates to the idea of normality (Goffman 1990: 4). Stigma is 'a characteristic of persons that is contrary to a norm of a social unit' (Stafford and Scott 1986: 80). Stigmatization comprises, and is representative of, the aberrant. The deviation from social normalcy then engenders an active rejection on that basis. Society constructs that which is (perceived to be) normal. Whatever trait fails to comply with these established expectations and social categorizations then becomes understood as 'not' normal. Deviation sets an actor apart as discredited or 'discreditable', to employ Goffman's (1990: 4) terminology. Interestingly, Richard Arneson (2007: 31) specifically describes the stigmatized person as socially regarded as 'disgusting'. Arneson's choice of phrase is related back here to the discussion of disgust above to demonstrate that repulsion is intrinsically related to the process of stigmatization.

Bruce Link and Jo Phelan (2001: 367–8) expand upon the idea of what we call the abnormal subject to argue that stigmatization is explicitly manifest as a form of labelling. 'Deviant', 'uncivilized', and 'odd'—are all potential labels that can be applied in order to other the stigmatized subject. The labels that communicate stigmatization are specifically associated with negative stereotypes and are expressed as the devaluation and dehumanization of the stigma. Applying certain identifiers enacts stigmatization in the form of social exclusion, distinction (such as an 'us v them' construction), loss of social status, and potential discrimination. Labels, therefore, provide both

The concept of stigmatization is now so extensive that it has been questioned whether it constitutes a sufficiently coherent and workable focus for analysis (e.g. Manzo 2004; Pescosolido and Martin 2015). Yet even diverse applications of stigmatization can still be conceptualized within the same terms of rejection on the basis of distinction from social norms.

[11] Goffman prefers the term 'social identity' because, in socially marking out another person, 'personal attributes such as "honesty" are involved, as well as structural ones, like "occupation"'. Goffman asserts that these identified features are not easily captured by the concept of 'status'.

the normative framing and also the language to other the deviant. Labels additionally reinforce the social belief at the heart of the stigmatization in question by engraining a given stigma's perceived validity.

The concept of a taboo, therefore, should be comprehended explicitly as a manifestation of stigmatization. What is taboo is marked out as abnormal in respect of societal expectation. In doing so, the taboo subject is one that has been constructed as beyond the tolerability of social standards and whereby the identification of that subject as deviant informs the normative response to it. Critically, this reference to stigmatization raises an important point about the issue of compliance with respect to the conceptualization of a taboo. Stigmatization would seem to be inherently connected to adherence. A stigma sets out standards of normality and so would appear to exist within the very context of its own non-violation. This observation could then be expanded to argue that compliance is in fact integral to the notion of a taboo. Indeed, the anthropological literature outlined above highlights that a key aspect of a taboo is the perception that there would be terrible consequences in breaking one. The same point is made by Tannenwald (2007: 11), who asserts that one feature of a taboo is the belief that non-compliance would have 'awful' and transformative implications. *This book* also openly describes a taboo as a type of prohibitive norm in that the study does not deny that a sense of prohibition is relevant. Yet there is an important distinction to be drawn here between a more general idea that a taboo should not be broken (not least because of the threat posed by the resultant consequences) and its actual breaking. The fears associated with violation are not the same as actual transgression in the sense of the debate highlighted above, whereby breaking a taboo is equated with its absence. The current understanding of a taboo is criticized here only for the way in which it dismisses a taboo on the basis that it is contravened. Consequently, debate can accept that a taboo is conceptualized in terms of the consequences of violation and still call it a taboo even if that violation occurs—in the same way that we would treat any norm with respect to stigmatization.

Bioweapons as the discreditable

Disease is stigmatized as a type of abnormality. Two core issues relate to this claim. First, biological infection questions the very notion and security of the self within the context of its perceived normalcy. Susan Miller (2004: 178) has said: 'Self-esteem is compromised by disease, as we begin to identify with dysfunction, ugliness, and lassitude.' Miller is effectively referring here to forms of abnormality. Disease places both ourselves and our bodies under a threat of the negative—such as the unattractive or becoming that which does not work. The negatives oppose and degrade what is deemed to be the normal. This argument reflects Menninghaus' (2003: 97) discussion of contamination as inconsistent with, and offensive to, the ideal of the 'perfect' body. In this case, factors such as ugliness would negate the identified ideal of the body itself. Miller's argument additionally mirrors Douglas' conception of a taboo as dirt (i.e. disease) being anomalous to the extent that it exists within the wrong place (i.e. the body). Drawing upon these perceptions of abnormality, disease is understood here as stigmatized in respect of its othering comparative to concepts of health. We value (what society considers to be) the healthy body as the normal. Disease represents the unhealthy and, therefore, the abnormal body. Sickness comprises an inversion of the biological relationship between health and the body to create an aberration and engenders feelings of repulsion as such.

The stigmatization associated with disease is even more keenly felt when the infection is deliberate, as with bioweapons (Barnaby 1997: 301; Cole 1997:13). Biological aggression comprises the very opposite of securing the healthy body and has been described as 'public health in reverse' on that basis (Greenberg 1968: 6). The sentiment essentially expresses the same belief inherent to the Hippocratic Oath. The oath calls upon doctors to reject any inappropriate and calculated harm to the body. Indeed, the classic version of the oath specifically prohibits causing harms of contamination, such as lethal poisoning by drugs (Hajar 2017).[12] The idea that a medical professional who is committed to preserving life would perform an act that undermines the healthy body is considered to be an anathema. In terms of biowarfare, the explicit removal of health on deliberate terms is similarly stigmatized as socially deviant. Analysis can take this point a step further by once again referring to Goffman. Goffman (1990: 11) considers how the ancient concept of stigma has now been expanded in respect of more contemporary social meanings. In doing so, he highlights that our wider concept of stigmatization has specifically become related to the attachment of stigma to 'physical disorder'. Any feature that deviates from the bodily norm is open to stigmatized construction. Disease then is integral and intrinsic to the very notion of stigmatization whereby infection: perverts our conception of the body

[12] The Hippocratic Oath has been modernized to address changes in moral concerns. The classic version prohibits proving lethal drugs and also to provide drugs to women to cause an abortion. These explicit commitments have been removed from the modern version in order to permit certain medical procedures, for example euthanasia and terminating a pregnancy (Hajar 2017).

itself; is constructed as an aberration from the bodily ideal; and is stigmatized accordingly on that basis.

Second, disease-as-a-weapon is stigmatized as taboo regarding an additional abnormality: a deviation from conventional violence. Neil Cooper (2011: 151) discusses the stigmatization of pariah weapons more generally, where he asserts that stigmatization is connected in part to 'the way their [pariah armaments] particular qualities are depicted as the antithesis of those possessed by legitimized and particularly heroic weapons'. Cooper's argument is expanded here to claim that the further those qualities are perceived to be from perceptions of conventional aggression, the more likely that the weapon in question is to be marked out as a taboo. Significantly, to talk of 'qualities' is necessarily a reference to social construction, even where this matter ultimately relates to the physical characteristics of a given weapon. Yet that construction still shapes a taboo and assists in explicating the modes of stigmatization inherent to it.

Relating this statement to biowarfare, biological armaments are differentiated from conventional modes of violence on the constructed basis of how they achieve destruction and the physical consequences inflicted. One aspect of this differentiation relates to the invisibility of disease. Symptoms of infection are often visible; indeed, that this visibility comprises an intrinsic stimulus to fear. For example, Ebola causes the victim to haemorrhage blood uncontrollably. Seeing or imagining these ghastly effects contributes to the associated horror concerning the sickness. Yet biological infection is also taboo in that disease itself is essentially invisible. Invisibility represents a core deviation from the conventional whereby this unseen-ness—and that, relatedly, disease is 'odorless, tasteless, silent and insidious'—will then 'instil a qualitatively different type of terror [to the conventional]' (Tucker 2001a: 264). Disease is also invisible to the point at which symptoms become seen. The incubation period between infection and the outward physical effects associated with a disease can involve a significant period of time. For example, Covid-19 has an incubation period of approximately five days. Consequently, it may take a week or more before an actor knows that they are infected with a disease. It can take even longer to establish that a full-blown attack has occurred within a given population. The time elapsed contributes to a differentiation between biological and conventional violence. W.H. Oldendorf (1962: 36) has demonstrated that a weapon which generates an immediate impact is more psychologically acceptable than one that more slowly takes effect and is 'quietly lethal' in this regard. Bioweapons exude stealth, clandestineness, and uncertainty. The identified characteristics strongly exacerbate perceptions of a taboo.

Biological contamination, moreover, is distinct from conventional forms of aggression in that disease is intrinsically invasive. Biowarfare comprises an internal attack upon the body. Consider why we find films such as *Alien* (1979) scary. The moment when the Alien bursts out of John Hurt's chest over an otherwise polite dinner is now iconic. The scene is especially shocking whereby it both exemplifies and amplifies our fear of being destroyed from within the very thing—our body—that keeps us alive and who we are. Compare the invasive nature of disease to a conventional explosive strike, say, in that the latter comprises an act of violence that occurs externally to the body. Admittedly, even acts of external violence are still deeply connected to the body. For example, the link between external violence and the body can be manifest in shifts in bodily perception undergone by disabled survivors who may experience their bodies very differently in terms of ability and changes to the bodily structure. Bioweapons, however, internalize harm to an even greater extent: to *within* the body. The invasive nature of disease additionally means that a victim cannot clearly escape a threat in the way they could run away from the enemy with the gun or evacuate the burning building (Stern 2002: 104). To the extent that the victim necessarily carries their contamination around within them, they cannot remove themselves from the zone of threat.

Contagion comprises a further point of deviation. Non-biological modes of weaponry cannot spread and replicate destruction in the way of communicable infection (although it is noted here that not all forms of biological agent are contagious, e.g. anthrax). The issue of contagion again connects to the ideas of excessive fear surrounding bioviolence (Mauroni 2000: 9). As Kellman (2007: 17) expresses it: 'Any other type of attack, regardless of its horror, is confineable in time and space; the harm is inflicted at the point of attack. . . . But a contagious bioattack somewhere puts everyone at risk everywhere.' The chain reaction impact of contagion also increases the mass destructive scope of a biological strike. The capacity to pass harm onto new victims means that the ultimate outcome of effect can be significantly disproportionate to that caused at the primary point of infection. Yet again, this deviation from conventional violence contributes to the construction of a bioweapons taboo as a type of abnormality.

Bioweapons are also perceived to be indiscriminate. The capacity of even targeted infection to spread beyond the battlefield means that biological aggression poses a distinct threat to non-combatants (Guillemin 2005: viii). On this issue it has been further highlighted that non-combatants are 'the weakest members of any nation at war' and that military actors would likely possess greater access to modes of biological protection, creating a disparity in how people may be affected in a biological strike (Clarke 1968: 6). There is pedigree in the claim concerning indiscriminate impact: indiscriminate effect is also cited with respect to WMD more generally as an intrinsic aspect of these armaments' taboo status and associated immorality (Stern 1999: 3, 12; Hashmi and Lee 2004: 9–10). Yet Tom Mangold and Jeff Goldberg (1999: xi) argue that bioweapons' capacity for contagion in achieving indiscriminate effect actually makes these armaments *more* immoral than their WMD counterparts. Significantly, the claim to indiscriminate destruction is imperfect. Susan Martin (2004: 33) demonstrates that bioweapons can be employed on a discriminate basis. Moreover, conventional armaments can also be used indiscriminately; for example, in the case of landmines which Asmeret Asefaw Behre (2005: 375) terms 'weapons of indiscriminate mass terror'. Consequently, this specific point of differentiation is subject to contestation. Yet the perception of indiscriminate force still remains key to explicating why bioweapons are constructed as taboo and comprises yet another way in which the conceptualization of disease-as-a-weapon is considered distinct as the basis of stigmatization.

Stigmatization as social control

In line with Koch, stigmatization not only comprises the identification of social aberration but also the prevention of deviance via 'social control' (Young 1928: 449). Protective control is enacted through the exclusion, and threat of exclusion, inherent to stigmatization. Specifically, control is achieved by: (1) reinforcing social expectations and providing the means to punish any deviation from them; (2) creating a pressure on society members to rigorously police those expectations whereby they too would be stigmatized if they did not demonstrate their own adherence in doing so (even if they had not performed a violation themselves); (3) inculcating feelings of shame in those who *do* deviate to encourage future compliance with, and provide an incentive to correct to, the normal; and (4) causing feelings of potential shame in those who *may* deviate, thereby acting as a deterrent against transgression (Goffman 1990).

The key theme in respect of stigmatization as social control is shame; explicitly, shame avoidance (Elster 1989: 99–100; M. Lewis 1995: 194–207; Pivetti, Camodeca, and Rapino 2016).[13] Shame delineates the boundaries of social normality and morality. Thomas Scheff (2003: 254) comments: 'Shame is our moral gyroscope.' Scheff (2000: 96–7) elsewhere talks of shame as the

[13] *The Scarlet Letter* specifically describes Prynne's 'A' as 'a mark of shame' (Hawthorne 1850).

expression of a threat to the social bond in that shame occurs when social ties are contravened. In order to avoid shame, therefore, those ties are secured and upheld. This concept of shame again relates to Link and Phelan's (2001) construct of labelling-as-stigmatization. The prospect of being tarred with a discreditable label incentivizes actor compliance with the norm. Labelling facilitates shame by demarcating the boundaries of the socially abnormal and creating a language of othering to be avoided, that is, the shame of being referred to in terms such as 'deviant'.

Stigmatization facilitates the social control related explicitly to taboos. Lynn Holden (2001: 12) says that taboos 'are symptomatic of a need for order ... without strict boundaries, it is feared the world will descend into chaos'. Evan Zuesse (1974: 486) similarly describes taboos as the 'ways of organizing the world around us into a ritualized social order, relieving uncertainty and directing behavior'. Taboos reflect an actor's desire for the social stability that the constructed normality appears to offer. Critically, the social control expressed within a taboo is self-reinforcing. Chaim Fershtman, Uri Gneezy, and Moshe Hoffman (2011: 140) refer to taboos as 'the thought police' because of the ways in which these norms condition actors as to what behaviours are deemed to be orthodox.

Stigmatization is discussed within IR. For example, Finnemore and Sikkink (1998: 892) assert in respect of international political behaviour: 'We recognize norm-breaking behavior because it generates disapproval or stigma.' Stigmatization inculcates international socialization and normative compliance. Actor behaviour is conditioned via the reinforcement of global norms whereby conformity 'meets the needs of social approval' (Shannon 2000: 300–1) and international legitimacy (Schimmelfennig 2000). Importantly, stigmatization has specifically been discussed in respect of the concept of taboo, albeit in the absence of a full discussion of disgust. In particular, stigmatization has been applied to the taboos concerning WMD (Shamai 2015), including nuclear weapons (Tannenwald 2007; Smetana 2019) and chemical armaments (Price 1997).

IR's discussion of stigmatization is expanded here with reference to reputation. Reputation constitutes a highly contested issue within the discipline. Analysts question whether reputation actually influences international politics, outcomes, and state behaviour[14]—and whether there even exists

[14] Some analysts argue that reputation does not influence international political behaviour (Huth and Russett 1984; Mercer 1996; Press 2007). Others argue that reputation does influence; for example, that reputation can influence political outcomes such as by making a state a more attractive ally (Sobel 1985; Chayes and Chayes 1995: 273; Gusinger and Smith 2002; Sartori 2002; Tomz 2007) and conferring greater power (Avdeyeva 2012: 300). In contrast, a poor or damaged reputation makes negotiation and international political engagement more difficult and engenders greater costs in doing so (Cowhey 1993: 300; Barrett 1999: 131, 139; Foot 2005: 293; Gibler 2008; Crescenzi et al. 2012). It is further argued states rely on

sufficient evidence to prove the matter either way (Gibler 2008: 426–7; Clare and Danilovic 2010: 861; Tingley and Walter 2011: 343; LeVeck and Narang 2017: 797). Yet it *is* maintained that state actors do care about their reputational standing (Levy 1997: 93; G. Miller 2003: 40; Wolford 2007; Clare and Danilovic 2010: 861; Meyers 2015: 19). Peter van Ham (2001) proposes the concept of the 'brand state' in which nations seek to construct and elaborate a certain international image. He argues that states curate their reputation as a form of political advertising on a metaphorical par with glossy travel brochures promoting their country as the ideal holiday destination. Van Ham demonstrates, therefore, that state actors do care how they are perceived on the global stage and will act in ways designed to promote that perception. As Thomas Schelling (2008: 124) famously argued, reputation is 'one of the few things worth fighting over' for state actors.

State actors' aspiration to maintain a positive reputation connects to stigmatization and societal control. Gregory Raymond (1997: 234) says of shame as an elicitor of social regulation: 'Similar processes also may operate among states belonging to an international society. Reputation and ostracism are important in the cultural context where national leaders regard themselves as part of something more than a system of states.' The desire to maintain a good reputation promotes an acknowledgement of the accepted norms and normality of the global political realm (Hurrell and Kingsbury 1992: 24; Haas 1998: 27; Guzman 2002; Wong 2016: 157). The issue also relates to multilateral institutions and regimes, as well as the treaties and agreements that uphold them. Legitimate membership of a global regime can convey a perception of actor prestige whereas non-involvement is associated with pariah status (Hafner-Burton and Montgomery 2006: 11). Stigmatization creates a poor international reputation; compliance, a credible one. Actors' need to maintain their reputation then explains why states respond to stigmatization as a form of control, as well as why taboos in particular drive their behaviour to such an extent.

International stigmatization is limited. Rebecca Adler-Nissen (2014) demonstrates that stigmatization is not an absolute means of social control and may, in fact, incite deviant behaviour. Ayse Zarakol (2011) explores the extent to which stigmatization and the pressures that it creates to conformity—especially where such pressure emanates from Western states towards the non-Western—may not engender genuine socialization. An actor may not truly inculcate a specific norm simply because they are

reputations in order to understand and respond to their international partners (Alt, Calvert, and Humes 1988; Kang 2004: 343; Crescenzi 2007: 382; Crescenzi, Kathman, and Long 2007).

pressurized to do so. Reputation may also mean that certain behaviours are dismissed as a fabrication. In the case of taboos, state actors may feign normative adherence to the extent that they wish to 'look good', but do not authentically internalize the taboo in question. The actor merely fakes compliance in a bid to maintain their standing, power, and political capacity (a point returned to in the next chapter). Furthermore, the desire to uphold a reputation may not lead to normative emulation but to concealment. Actors will still violate a taboo—yet they will endeavour to not get caught. Covert behaviour makes norm regulation harder as transgression becomes more difficult to detect (Olsthoorn 2008: 472). Despite these caveats, however, stigmatization remains a core factor in the institutional and social function of the international community, of which taboos are a part. Stigma is integral to the collective survival of that community via the normative adherence to taboos.

Controlling bioviolence

Disease is intrinsic to conceptions of society (Rosenberg 1986). Sickness comprises 'one of the most fundamental factors in human affairs' and 'the manner in which people respond to disease can illuminate the underlying values that hold a society together and give it a sense of coherence' (Grob 1977: 396; see also Anderson 1999: 245). Consequently, disease is intimately connected to ideas of social stability. The connection overlaps here with the function of a taboo in seeking to prevent both physical and moral contagion; indeed, many taboos express social rules explicitly concerned with precluding disease.[15] Controlling disease, therefore, relates to the control of society.

A parallel argument exists within the literature on biopolitics. Biopolitics now comprises an extensive and contested field of analysis which is considered 'legion' within various sub-disciplines of IR (Cavanagh 2014: 273). Debate, however, often relates to Michel Foucault (2004). Foucault discusses biopolitics as the relationship between power and life; more specifically, the power *over* life. This relationship is not simply the idea that power determines life in the sense that a subject then lives or dies. Biopower 'is the power to "make" live and "let" die' (Foucault 2003a: 240). Foucault's work then expresses 'a modern form of governance that conceives biological life

[15] The claim here holds even where the identified restraints may no longer have relevance in protecting modern societies, i.e. the taboo prevents against a type of disease that has since ceased to constitute a threat to a population (Weiss, Ramakrishna, and Somma 2006; R.A. Smith and Hughes 2014).

as a resource' (Greco 2019: 110). Within this theoretical context, Foucault (2003b: 188) additionally discusses the idea of the 'pathological life' in which he claims that disease is not simply an external threat to our existence, but a case whereby life and disease are intrinsically 'bound together'. On bioweapons specifically, Emilio Modini (2005: 253) has said that 'biowarfare is the ultimate form assumed by biopolitics'.

Subhadeepta Ray (2021: 16) applies Foucault's conception of biopolitics to interpret viruses as a form of disease governance (see also Fuller 2016; Kong 2019). Political responses to illness are not socially neutral. Bruce Braun (2013: 45) describes the situation as one wherein 'biosecurity practices bring life into the realm of political calculation'. Policies pertaining to *how* a disease is treated and *who* receives treatment engender social implications. Foucault (1995: 198) argued that lepers, for example, 'gave rise to rituals of exclusion' through which a disciplinary power was exerted to control them. Controls not only relate to how infected persons are addressed, however, but also the measures of preparedness which are introduced (or not) to preclude infection (Sanford, Polzer, and McDonough 2016). In both cases, the stigmatization of those actors who become, or hypothetically can become, ill determines their social identity. The othering of the sick persons from society comprises an intrinsic aspect of this process. As already discussed above, the diseased body is considered distinct and engenders policies that both express and construct that body's deviation from the normal life, that is, health. The society in question is not just national but can also be international in scope (Bashford 2006; Ingram 2009).

Social prohibitions relating to disease can be employed in respect to highly negative means of control, including racist discrimination (Sontag 1978; S. Cohen 2012). Cholera has been demonized as a specifically 'Asian' disease dating back to the 1800s, which Christopher Hamlin (2012) terms the 'cholera stigma'. Jews were routinely accused of spreading plague (Cowdrey 1984: 154). Covid-19 has underpinned an increase in racial prejudice and hate attacks due to the virus' discriminatory construction as a 'Chinese' sickness (Addo 2020). The matter is not limited to cases of actual disease, however, but also relates to the metaphorical. Rowan Savage (2007) demonstrates that the targets of genocide may be othered on the basis that they are associated with disease as a metaphorical expression of that which is not pure. Consequently, it is argued that racist social constructions of biological infection lead to a biopolitics of 'differential vulnerability' by which a disease impacts disproportionately upon societal members (Lorenzini 2021: S43–S44). The same argument is applicable to other forms of discrimination such as the detrimental associations between disease and sexuality; notably, the negative construction of HIV/AIDS in respect of the gay community (Leiker, Taub, and Gast 1995).

The stigmatization of disease as a form of social control links to the bioweapons taboo whereby the taboo also seeks to preclude biological infection. The taboo comprises a normative response constructed to keep disease, and therefore social stability, in check. Stigmatization further pertains to the pressure to conform with the taboo itself. Biowarfare is constructed as a social wrong that must not be transgressed—else the transgressor would themselves be stigmatized and labelled as having violated the expectations of civilized conduct. Biowarfare is 'described as barbarous, a morally repugnant form of warfare, a "dirty" way to fight' (Mauroni 2000: 4). Bioviolence is viewed as 'unchivalrous and beyond the pale of civilised warfare' (T. Cook 2000: 47). Bioweapons are 'totally evil, degrading the better instincts of humanity' and 'they contravene an essential value among civilized people' (SIPRI 1971a: 79; Cole 1997: 13). Paul Cassell (1983: 260) further relates this stance against biowarfare to international opinion and response. Cassell asserts that 'when one nation accuses another of using these weapons, it is understood that the allegation carries with it a call for moral condemnation far in excess of that which would be called for in response to more conventional warfare'. Furthermore, the taboo is also codified as a type of social control. This book will outline multiple international agreements which uphold the taboo's normative expectations including the Geneva Protocol and the BWC. The documents reinforce the stigmatization surrounding bioweapons whereby they establish official standards of behaviour. Biowarfare has then been socially institutionalized as a taboo form of aggression to the extent that the prohibition shapes behaviour, actor judgement concerning such behaviour, and social response (including codification). The construction of biological armaments as abnormal to acceptable social behaviour sets certain expectations for how life itself should be treated in respect to the threat of biological violence. In the case of biowarfare, these social expectations have established that these weapons are off-limits.

Fetishization

It has already been noted that there exist two key elements to a taboo: the unclean and the sacred. Stigmatization is effectively used within IR to explicate the former, that is, stigma as the socially deviant and discredited. What of the latter? IR scholars do acknowledge that the sacred is elemental to the anthropological aspect of taboo (e.g. Gizewski 1996: 399) and the connection between taboos and what is sacred is discussed by the likes of Philip Tetlock (2003). IR analysis, however, remains undeveloped on this point. In particular, IR lacks the academic tools and framework necessary to fully explain what dynamics cause a taboo to stand out from more generalized notions of denial. Furthermore, the discipline cannot comprehend other, more complex, and often desire-driven, notions of disgust also expressed as a taboo. This section addresses the significant gap in analysis to argue that a further concept is required in order to model the character and social application of taboo: *fetishization*.

Fetishization concerns the reification of a subject, also known as a fetish. The fetish was traditionally understood as a physical item—but can now also include the non-physical, for example a state of thought (Ellen 1988: 232), abstraction (Silva 2013: 80), or 'mode of action' (Johnson 2000: 249). Peter Pels (1998: 91) refers to the spirit 'of' as opposed to the spirit 'in' matter whereby the physical presence of an object is now unnecessary in identifying a fetish.[16] The process of reification is difficult to define precisely (Pitkin 1987). Yet fetishization essentially comprises a case whereby a subject comes to possess an exaggerated social value, including an explicitly sacred one (B. Smith 1995: 1170; A. Reed 2013: 152). The value is specifically discussed as a 'false value' (Keane 1997: 677; Petrof 2015: 217). Indeed, fetish is an English word adapted from the Portuguese *feitico*, meaning 'made' or 'artificial' (Nassau 1904: 262). Theoretically, therefore, without its fetishized value the subject would simply hold its 'normal' worth. The value is also self-reinforcing. Consequently, there will always be a question as to whether a taboo exists because it is fetishized or whether it is fetishized because it is a taboo, where the concept of taboo itself possesses a value. Fetishization, however, remains intrinsic to a taboo and a fundamental aspect of its construction as such.

Three key strands of fetishization are identified (Ellen 1988). First, (as with the concept of taboo) fetishization originated as an anthropological construct. A fetish was a revered object thought to embody a sacred and supernatural element within a 'primitive' culture (Lang 1879; Pool 1990; Ampuja 2020: 32). This line of thought fell out of academic favour in the mid–late nineteenth century—largely thanks to a rejection of the idea that societies could be considered 'primitive' (Ellen 1988: 214–15; Graeber 2005: 410). Tomoko Masuzawa (2000: 242) says: 'Even by Victorian standards ... the notion of fetishism was already embarrassingly outmoded, something

[16] Fetishization has even been applied to IR theory itself. Hedley Bull (1966: 372) employs the concept to criticize the scientific method in IR and claims that this approach 'is in some cases distorted and impoverished by a fetish for measurement'.

rather more reminiscent of certain older habits of thought than a critical tool of the emerging scientific discipline.' For example, William Robertson-Smith (1884: 209) dismissed the concept of fetishization on the grounds this now comprised 'merely a popular term, which conveys no precise ideas, but is vaguely supposed to mean something very savage and contemptible'.

Fetishization re-emerged in somewhat rehabilitated form within the work of Freud and Karl Marx—the two other strands. Freud's (2001a) fetishization is inherently sexual. Freud's theoretical framework centres on the penis; explicitly, the castrated penis of the mother. Once a young male realizes that the mother has no penis he becomes concerned by that absence and experiences a further anguish that he himself is at risk of castration. The lack of the penis is expressed in the identification of sexual fetishes as a form of substitute, for example feet, velvet, and fur. The penis is 'replaced' by other objects, which are then attributed an extreme value as such. Freud (2001b: 153) specifically harks back to the anthropological understanding of the fetish in constructing his theory, whereby he states: 'Such substitutes [for the penis] are with some justice likened to the fetishes in which savages believe that their gods are embodied.' Freud effectively expresses the same concept of a revered and excessive value which is still understood as elemental to the idea of fetishization itself.

Freud's theory lies in contrast to Karl Marx's (1909: 62–9) commodity fetish. Commodity fetishization occurs when commercial goods become associated with an inflated sense of worth as a consequence of the production system. Like Freud, Marx's theory also has its ideological roots within the anthropological literature. Marx adapted his term from Charles de Brosses' 1760 prominent text on religion and idolatry, *Du Culte Des Dieux Fetiches, Ou Parallele De L'ancienne Religion de L'Egypte Avec La Religion Actuelle De Nigritie* (Boer 2010: 96). Significantly, the fetishized value in question is not connected to the commodity itself, but the social relations that construct it—what Marx (1909: 63) famously conceptualizes as the 'fantastic form of a relation between things'. Marx accepts that people understand this value exists only as the consequence of the identified social relations. They act, however, as if the value *was* intrinsic—whereby this perception functions as a type of recognized misrecognition.[17] Slavoj Zizek (1989: 31) comments that we 'are fetishists in practice, not in theory'. Consequently, Marx's theoretical model highlights that fetishization is not merely a form of reification,

[17] In a similar vein, Freud (2001a: 157) asserted that the fetishist is able to both believe in a sexual fantasy and at the same time accept that it *is* a fantasy. Moreover, the mother is simultaneously understood to be both not castrated and also treated as though this were genuinely the case.

but a specific form of reification that relates to, and emerges from, social construction and interaction (Comor 2017: 7).

All three core conceptualizations now underpin more general analyses based on fetishization. The strands are used often in combination and are applied to issues other than those overtly connected to anthropological, sexual, or commodity desires (Pietz 1985; Pool 1990: 122–4; Pencheva and Maronitis 2018: 529). Crucially, the academic discourse does not consider itself bound or biased by the concept's discriminatory genesis within anthropology and its early construction of the primitive—what has been called its 'sinister pedigree' (Pietz 1985: 5). It is argued that modern analysis can overcome this 'dubious history' and is able to 'negotiate the term's colonial heritage' in ways that overcome this difficult baggage (Wark and Wark 2019: 296).

Fetishization as a socially constructed, exaggerated, and contrived form of worth is key to a taboo. The perception of value connects to disgust and stigmatization, especially where the breaking of a taboo involves that value's violation. Yet value also functions within its own right as a necessary characteristic of taboo. In particular, it explains why we respond with greater fervour to a taboo than other forms of prohibitive norm, that is, we attach a higher value to it. Moreover, the identification of the sacred also relates to the emergence of a taboo. Douglas Marshall (2010: 67–8) argues that the sacred is produced when temptation clashes with social expectation (to not engage in the temptation) and whereby the discrepancy between the two is 'rationalized' by denoting the relevant subject as sacred. The identification of the sacred in this manner then translates into a taboo. Fetishization further explicates taboos that involve an explicit desecration of the sacred in the more traditional sense. Swearing in a church is not merely swearing; although swearing itself may comprise a taboo depending on the severity and sheer creativity of the expletives in question (Jay 2009). The act is also blasphemy, wherein the sacred comprises an integral aspect of its taboo nature. Disgust and stigmatization alone cannot fully conceptualize why the violation is abhorrent in that the church itself comprises a taboo underpinned by a rejection derived from reverence.

Fetishization additionally models another aspect of taboo that stigmatization cannot. It expresses the ways the disgusting can—to adopt William Miller's (1997: x) phrasing—'attract as well as repel'. At one level, this sensation is manifest as a compelling, if possibly unwelcome, intrigue as to what disgusts. Even to the extent that we are repulsed by the disgusting, 'it rarely does so without also capturing our attention. … We find it hard not to sneak a second look' (Miller 1997: x). Plato (1987) recalls the tale of Leontion in

The Republic. Leontion encounters some corpses recently dispatched by the public executioner and he cannot help but stare:

> He [Leontion] wanted to go and look at them [the dead], and yet at the same time held himself back in disgust. For a time he struggled with himself and covered his eyes, but at last his desire got the better of him and he ran up to the corpses, opening his eyes wide and saying to them, 'There you are, curse you—a lovely sight! Have a real good look!'
>
> **(Plato 1987: IV 439e–440a)**

At a further level, disgust may comprise an even deeper expression of desire. As Marcus Tullius Cicero (1942: 81) asserted: 'Thus in all things the greatest pleasures are only narrowly separated from disgust.' Freud again constitutes an ideal citation whereby a significant aspect of his analysis explores the sexual attraction to that which would typically be deemed repulsive. In a similar vein, Aurel Kolnai (2004b: 102) talks of *haut gout* or high flavour; for example, the 'horrible' smell of strong cheese or game, but where these foods are considered superior for being more pungent. (Although it should also be noted that Kolnai (2004a: 63) states elsewhere that one function of disgust is to prevent us 'drowning in pleasure' by placing limits on our capacity for indulgence.) Critically then, actors may desire what they understand to be disgustingly taboo. Indeed, they want the subject precisely because it is taboo—albeit where that desire still conflicts with a continued and deeply held fear associated with transgression (O'Doherty 1960: 132).

It is easy to reduce a taboo to absolute negative rejection: thou shalt not, in the vein of disgust and stigmatization. Introducing fetishization into the analytic framework, however, highlights that disgust-as-taboo is experienced in more complex and often desire-driven ways. Taboos express not merely repulsion, but also captivation as well as a deep sense of urgency and prioritization. Such types of response cannot be fully captured within a model of taboo not supplemented by the additional incorporation of fetishization as a core analytic factor. It is only when these three characteristics are applied in tandem that we can know when we are dealing with a taboo. Importantly, these factors will not always apply to the same degree in respect of every taboo. Each taboo is distinct and manifests within our thinking to differing extents. Yet these three characteristics remain, in combination, a necessary condition for the identification of taboo—explicitly where this underpins the identification of taboo *as something distinct* from a non-taboo norm. In cases where a prohibitive norm is perceived to be disgusting (that we experience revulsion towards the subject of the norm), stigmatized (socially rejected), and fetishized (particularly in the sense that we attribute a higher sense of value and priority), this lays the ground for the recognition of a taboo. In particular, IR cannot effectively rely on stigmatization alone to analyse taboo and requires reference to this full range of characteristics to understand the identification and function of a taboo.

Burned up with fire

> [I]n one day her plagues will come—pestilence and mourning and famine—and she will be burned up with fire; for the Lord God who judges her is strong.
>
> **(Revelation 18:8, The Bible 1989)**

Disease has long been constructed within the context of the sacred. There is little that is more Biblical than a plague. Ask Exodus. Yet the sacred perception of disease is not limited to Christianity. The 'savages' of the anthropological literature frequently considered biological infection to be connected to 'spiritual or supernatural' forces, where this understanding underpinned conceptions of a taboo ('Medicine', 1915: 752). Ancient cultures also viewed disease as both a natural cause and a 'supernatural punishment' (Sontag 1978: 43).

A more general link between fetishization and weaponry is also long established (e.g. Salter 2014: 163; Bentley 2018). Fetishization is especially applicable to biowarfare, however, given the disgusted and stigmatized sentiments pertaining to disease already outlined. In particular, fetishization relates to existing claims that biowarfare is constructed within disproportionate terms. Bioweapons' disproportionality connects to perceptions of the excessive value of fetishization and is evident in the ways that bioviolence can 'arouse a *peculiar degree* of ire and passion [emphasis added]' (Spiers 2010: 11). Praful Bidwai (1999: 109, 110) identifies a 'unique mystique' about WMD which is derived from 'a special fascination' with their destructive power but where that construction is not necessarily proportional to what those weapons can achieve in terms of physical effect.

Comparison can also be drawn with the work of Anne Harrington de Santana (2009), who applies a Marxian conception of commodity fetishism to nuclear weapons. She describes nuclear armaments in terms of their 'fetishism of force'.[18] Harrington de Santana (2009: 325) concludes that

[18] Harrington de Santana's (2009: 339) work also plays into Freud where she describes nuclear weapons as 'sensuous things'—although she does not mention Freud explicitly.

'nuclear weapons function as the currency of power in the international system' on this basis. Like Bidwai, she dismisses the idea that nuclear weapons are prioritized as a consequence only of their 'capacity for destruction' (Harrington de Santana 2009: 327). Harrington de Santana demonstrates instead that it is fetishization which explains the greater value placed upon this mode of aggression. The same argument is extended here to bioweapons as yet another form of pariah armament. The disproportionate reaction to both the extent and nature of bioweapons' effect underpins the way in which these devices are fetishized and subsequently constructed as a taboo.

Fetishization of biowarfare additionally relates to the concept of disgust-as-desire. As Jez Littlewood (2007a: 193) states: 'Biological weapons are exotic.' Bioweapons inspire not only fear but also intrigue on a scale that is not associated with conventional modes of violence. For example, there is good reason why Hollywood films are replete with tales of devastating pandemics (Tomes 2002: 625; Dehority 2020). Pandemic films cause and play on feelings of intense anxiety. The films depict mass death, gruesome suffering, and social collapse (Han and Curtis 2020); yet the shocking fear associated with fictional portrayals of out-of-control biological destruction is also sufficiently attractive to sell numerous movie tickets. Disease fascinates us at the same time as it horrifies. Indeed, disease fascinates *because* it horrifies. In the same way that Leontion could not help but gaze upon the rotting corpses in the street so our interest is pricked at the idea of using disease as weaponry. Via fetishization, therefore, bioweapons have been widely established as distinct in respect to political thinking. More specifically, the distinct nature of biological armaments as possessing an excessive value in our minds is manifest within the concept of the taboo.

A new model of taboo

We require a new conceptual model of taboo. Current definitions as to what a taboo is are questioned here on two key grounds. First, the existing analysis has yet to fully articulate the normative characteristics of a taboo. Second, the debate on taboos is further shown to be overly dependent on compliance as the primary analytic standard for a taboo's identification. Dependence is both explicit and implicit. Responding to these critical gaps in our analytic understanding, this chapter has proposed a more comprehensive and effective means of comprehending the concept of taboo. The new model identifies a taboo as a prohibitive norm which: *disgusts*, is *stigmatized*, is *fetishized*. If a norm fulfils these necessary criteria, then it can be considered a taboo.

The extent to which the characteristics of the new model still relate to compliance should not be overlooked. Taboos are dangerous—and that danger then comprises a basis of, and desire for, avoidance. The model, however, makes two key contributions to the current understanding of compliance as a primary measure of taboo. First, the model comprehensively articulates the normative characteristics of danger associated with taboo, where this formulates the criteria for identification. Second, the model highlights that compliance does not comprise the be-all-and-end-all of that identification. To the extent that some scholars do not explicitly adopt compliance in the vein discussed in the second point, the contribution may only relate to the first. Even on this basis, however, the model still makes a critical addition to the debate by now clearly establishing the terms on which a taboo is seen to exist, which has previously been lacking.

Biowarfare meets this new definitional standard of a taboo. Bioweapons are disgusting in that they are intrinsically repellent. Repulsion is not only the consequence of the weapons' potentiality for mass destructive effect. Disgust also elicits from an abhorrence of disease as an abysmal, invisible, and indiscriminate contaminant. Furthermore, disgust is not limited to ideas of biological contamination, but additionally represents the rejection of moral infection in that these very weapons are deemed unethical. Bioweapons are stigmatized whereby the disease they spread is opposed as the emotive inverse of the normal and uncontaminated body. Bioweapons are further segregated and controlled on the basis of their constructed deviance from modes of conventional aggression. The same features that make bioviolence disgusting (invisibility, invasiveness, infectiousness, and indiscriminate impact) are similarly the grounds upon which these weapons are differentiated from the conventional—and by which the biowarfare threat is then subject to stigmatization and rejection. Finally, bioweapons are fetishized—particularly in the sense that they are attributed an excess value. Biological armaments are not ordinary weapons in our mind. More specifically, these weapons are not ordinary as a consequence of, and contribution to, their status as a taboo. In terms of what is reified, what is sacred, and what titillates us—bioweapons stand out as distinct. These armaments stand out as taboo.

2
Measuring the Taboo

Having established what a taboo is, this study now turns to the questions concerning whether taboos influence actor behaviour within the international realm and how analysis can measure such influence. The argument presented here maintains that taboos do shape actor behaviour to a decisive extent in respect of international politics. Yet this claim is not without controversy. Scholars contest the ultimate significance of norms within IR—and on security matters in particular. More specifically, it has been argued that actors are prepared to sacrifice norms to rational need and self-interest when pressed, regardless of however much they may desire to conform to the normative standard. The theoretical dissonance between these two positions can be outlined in terms of the conflict between James Marsh and Johan Olsen's (1998) logic of consequences (actor self-interest) and logic of appropriateness (normative adherence). While upholding the relevance of taboos essentially prioritizes the logic of appropriateness, it is claimed elsewhere that—should that logic fall at variance with the logic of consequences—then it is the latter that will prevail. As the former UK Foreign Secretary, Jack Straw (in Black 2003), once opined: 'When rules clash with reality, reality normally wins.' Stephen Krasner (1999) discusses the same idea within his analysis of organized hypocrisy:

> Rulers are calculators, not agents manifesting some deeper international institutional structure. ... Rulers may appeal to norms that do depend on intersubjective shared understandings, albeit understandings that can be manipulated, but they may also appeal to guns, and a bullet through the head has the same effect regardless of the cognitive perspective of the target.
>
> **(Krasner 1999: 50–1)**

Jack Goldsmith and Eric Posner (2002) take this argument further to assert that actors will only be seen to adhere to normative and moral expectations when those standards conveniently align with, and assist in the realization of,

The Biological Weapons Taboo. Michelle Bentley, Oxford University Press. © Michelle Bentley (2023).
DOI: 10.1093/oso/9780198892151.003.0003

self-interest. Consequently, George Ignatieff (1974: 92) concludes of international policymaking that actor behaviour has 'more to do with horse trading than with St Paul'. His claim is that self-interest ultimately trumps moral sentiment and normative belief, including taboos.

In demonstrating here why this line of argument does not in fact dispute the significant (in some cases, ultimate) impact of taboos, the logic of consequences should not be applied to set up a straw person argument against the claim to a taboo. It is not the case that we can frame the IR discipline as a binary between those who oppose normative influence on one side and those who accept norms as an overriding factor on the other. Such an accusation of misstating disciplinary understandings of normative influence in this way has already been levelled against Tannenwald and her seminal discussion of the nuclear taboo. Tannenwald's analysis presents the taboo as a corrective to what she interprets as a realist interpretation of the non-use of nuclear weapons, where this is constructed as a logic of consequences. Tannenwald (2007: 4) equates realism with a norms-do-not-matter claim: 'Realists would deny that a taboo exists or that, at minimum, it can be usefully distinguished from either the material interests of the actors or the behavioral pattern of non-use.' Within Tannenwald's reading, realists consider norms to comprise merely 'the frosting on the cake' of international politics (Tannenwald 2007: 39). Tannenwald subsequently seeks to dispute the validity of what she identifies as the realist position as a means to then uphold the taboo as a major influence in respect of nuclear politics. Tannenwald's refutation of realism, however, has been strongly rejected as an empty suit argument that misunderstands both realism and ignores a wider disciplinary recognition that norms do play a role, even if the extent of that role is subject to contestation. Janne Nolan (2008: 540) pulls no punches on this point: 'Tannenwald spends too much time bashing overly simplified caricatures of realism and deterrence that seem unworthy of the effort.'

Despite these criticisms of Tannenwald's interpretation and position, the distinction between the two logics can still be applied to open up key questions concerning the extent of a taboo's influence, especially vis-à-vis dynamics that would seem to reflect the logic of consequences. In doing so, Tannenwald's wider approach still provides an effective means of establishing that taboos do influence actor behaviour in two valuable ways, which are extended here to the study of bioweapons. First, Tannenwald delivers an important understanding as to how taboos function. Her analysis is employed within this study not only to ascertain what taboos 'do' but also to stress the substantial relevance of taboos—in line with this book's assertion that such norms decisively impact upon decision-making. In analysing Tannenwald's contribution, however, this study contextualizes her claims—and also later the work of Price—in terms of the new model of taboo proposed within the previous chapter. The model's problematization of normative compliance has conceptual and methodological implications which relate to the identification and measurement of a taboo's effect. This chapter explains these implications and what they mean for the study of international taboos. Second, Tannenwald provides a methodology for the empirical analysis of taboos based on the concept of 'taboo talk'. The chapter outlines Tannenwald's method and develops the approach further in connection with the concept of strategic narratives.

The relevance of taboos

Tannenwald upholds the core claim that taboos matter within the international political realm. Her theory builds on a much wider literature on norms within the IR discipline which accepts that normative expectations are 'a force to reckon with' in respect of international behaviour (Jacobson 2008: 360). Tannenwald cites key scholars such as Alexander Wendt (1999) and Katzenstein (1996) to place her work within a constructivist tradition that considers norms to constitute a primary factor within IR. International politics 'is only possible within an intersubjective social context' and, therefore, without norms 'exercises of power, or actions, would be devoid of meaning' (Hopf 1998: 173). Norms are not merely a secondary consideration that can easily fall prey to material concerns and self-interest. Norms constitute a fundamental and intrinsic aspect of international decision-making, of which taboos represent a particularly notable example.

Tannenwald explains how taboos influence actor behaviour. Tannenwald (2007: 44–7) demonstrates that taboos exert three core normative effects. First, taboos express a regulative effect whereby norms constrain actor behaviour. Taboos cause actors to conform to a normative standard. A key point here is the idea that these certain types of norm explicitly restrain. The transgressive behaviour could happen, but the taboo precludes this from occurring; that is, the normative expectation on the actor is intrinsically prohibitive. Tannenwald (2007: 44) explicitly states that a taboo 'constrains a behaviour (nuclear use) that would exist whether or not there were rules about it'.

Second, taboos are normatively constitutive. Taboos shape actions within the international system by ascertaining what types of behaviour are deemed permissible and creating the 'fundamental categories though which actors conceptualize the world' (Tannenwald 2007: 45). From a constructivist perspective, the constitutive impact of taboos explains how interests (which the realist position may reduce to self-interest) are the product of normative expectation and its relationship to the validation of actor identity. An example of the constitutive effect would be the WMD categorization. The term identifies certain weapons not only as enabling mass destruction in terms of physical effect, but also expresses the normative belief that these weapons should be prohibited whereby they are then deemed to be pariah. Taboos embed specific assumptions and practices which determine what is and what is not acceptable as behaviour—and explicitly what types of conduct would be anticipated of a 'civilised' state (Tannenwald 2007: 46).

Third, taboos are permissive in that they cause secondary, and perhaps unintentional, effects. Tannenwald (2007) explains the concept of these other effects in more detail:

> This refers to the way norms—taboos in particular—by serving as focal points, selectively divert our normative gaze. By categorizing weapons in certain ways … norms and taboos may obscure other 'facts' about the world and shield other practices from attention. Thus one permissive effect of the nuclear taboo may be to shield non-nuclear weapons from normative opprobrium.
>
> **(Tannenwald 2007: 46–7)**

Permissive effects may not be deliberate or even desirable. These effects, however, still indicate the presence and influence of a taboo.

The identified dynamics are maintained via three primary mechanisms: 'domestic opinion, world opinion, and the personal moral convictions of individual leaders' (Tannenwald 2007: 47). Domestic public actors exert influence over their leaders and representatives if only because those political figures wish to be re-elected (Tannenwald 2007: 48). The direct impact of a domestic audience on foreign policy in particular, and whether that audience even should have an impact, is contested (see Aldrich et al. 2006 for an overview of the debate). Yet the factors of globalization, the politicization of foreign policy, an increasing interest in global affairs at the level of public opinion, and a growing need for states to establish their legitimacy in respect to international politics are all issues that have been cited to challenge this viewpoint (Zerka 2021). Domestic public opinion is then highlighted as a key—if not always decisive—pathway of a taboo. This book will confirm Tannenwald's conceptual approach in respect of the public impact on biowarfare policy.

Taboos are also facilitated via world opinion. Tannenwald (2007: 49) disputes the claim that world opinion simply comprises a reflection of the dominant states' own interests. In evidence of this position, she identifies the presence of a clear international opinion manifest within the construction of global institutions that give a collective voice to all state participants of the global political system. With particular reference to American dominance of that system, she notes that 'the values and norms promoted by the UN have often gone against the preferences of the United States' (Tannenwald 2007: 49). State leaders—even within the US—accept and respond to the expectations set by the wider international community and are held accountable by that same community.

Finally, political actors may also express their own personal convictions concerning the applicability of a taboo. Tannenwald (2007: 50) does note that those convictions may actually comprise the product of the first two pathways to the taboo. Yet her initial point is still a valid one: individual views matter. The taboo can be realized because a particular actor believes, or is advised by someone who believes, in the norm. The issue is especially pronounced when applied to actors in a specific position of power, such as the US president. Overall then, Tannenwald concludes on the basis of this model that taboos exert a significant impact on actor behaviour, where that behaviour then aligns with the expectations of these taboos.

Alternatives to the taboo

Tannenwald seeks to validate the presence of a taboo pertaining to the first-use of nuclear weapons in two primary ways: proving its existence and by disputing alternative explanations for non-employment. Tannenwald (2007: 53) equates these alternatives with realist interpretations and terms them 'non-norms' factors in order to reflect the ways in which these other explanations disregard normative influences by her reasoning. Tannenwald (2007: 30) demonstrates that actor aversion to first-use nuclear weapons employment is typically reduced to one of five 'non-norms' factors: '(1) deterrence; (2) the undesirable or uncertain long-term consequences of a use of nuclear weapons; (3) nuclear weapons' lack of military utility; (4) non-deterrence practical considerations i.e. bureaucratic and readiness factors; and (5) the obsolescence of war'. Tannenwald provides a criticism of each alternative and then applies this refutation to propose the taboo as a 'sixth' factor in explicating nuclear weapons' non-use. Realist interpretations are inadequate; ergo, the taboo can be brought forward as an explanation for actor behaviour in respect to no-first-use.

Tannenwald's five factors parallel arguments that can also be identified in respect of the debates concerning biowarfare. The evidence in this book demonstrates that actors have articulated the following claims pertaining to biological violence: (1) bioweapon use is deterred by the threat of like-for-like retaliation; (2) bioweapons should not be employed whereby they create uncontrollable and indiscriminate outbreaks of disease; (3) bioweapons lack military utility; (4) bioweapons are poorly developed, thereby creating a readiness issue; and (5) bioweapons are too destructive with respect to modern conflict. Critically, the application here of Tannenwald's analysis to biowarfare is not to replicate her reduction of these factors to her proposed conception of realism, per Nolan's criticism.[1] While these identified alternatives frequently conflict with what would be understood as an absolute manifestation of the taboo, it would be inaccurate to equate these arguments to an entirely 'non-norms' position. The realism issue aside, however, Tannenwald's identification of other explanations still helps to identify key claims surrounding biowarfare of relevance for this study, at least at a very general level. More specifically, that—as Tannenwald argues in respect of the debates on nuclear weapons—there also exist comparable problems with the above alternatives relating to bioweapons which it is useful to cite here.

A key issue with our analysis of biowarfare is that it is simply not known whether bioweapons exhibit sufficient militarily utility. The lack of bioweapon use translates into a lack of quantifiable knowledge as to what these armaments can achieve in terms of strategic and tactical effects. Furthermore, the capacity for testing is too limited to resolve the matter (Steinbruner 1997: 88). Testing on a scale designed to ascertain the full effect of bioweapons could potentially cause a major disease outbreak or would involve a level of human experimentation that is not ethically permissible. Consequently, analysts, policymakers, and military actors may speculate upon the utility of bioweapons (and, as this book will show, they do so frequently) but there exists no demonstrable consensus.

The uncertainty concerning bioweapons' effect has engendered a highly convoluted debate on this matter. On the one hand, bioweapons are deemed to be militarily effective. Bioweapons are 'unfortunately characterized' by being difficult to detect (and, therefore, also to prevent against or mitigate) where: they comprise slow-acting armaments; outbreaks could be misconstrued as natural infection; and attribution is problematic (Danzig and

[1] This book does refer to material factors and self-interest, often as dynamics that are in conflict specifically with the bioweapons taboo. It is noted, however, that this type of reference needs to take into account the wider disciplinary relevance of norms and also constructivist claims that issues such as 'interests' can be constructed in relation to norms and identity (Risse and Wiener 1999: 778–9).

Berkowsky 2001: 9–10). Bioweapons are also associated with the following traits: high accessibility (agents are often present in nature); cheapness; relative ease of delivery, requiring only small amounts of agent to cause potent effect; involving processes of weaponization that can be easily concealed within dual use research; and generating mass panic (Danzig and Berkowsky 2001: 9–10). All these features could be seen as making bioweapons appear effective and attractive. Furthermore, bioweapons are recognized as a type of WMD. Biological armaments are seen to possess a capability for devastating destruction on the scale of an epidemic or pandemic. Analysts argue in this vein that bioweapons are equal to, or even more deadly than, nuclear arms (Betts 1998: 32): 'Biotechnology is one of only two technologies [biological and nuclear] that truly deserve the label *agents of mass destruction*' [emphasis in original] (Steinbruner and Okutani 2004: 273). Bioweapons' capacity for extreme effect can be highlighted as evidence that these armaments do possess utility. This observation, in turn, would undermine the claim that aversion to bioweapons is attributable to a lack of military effectiveness—as Tannenwald (2007: 39–42) argues in respect of nuclear armaments.

On the other hand, however, analysts do still dispute the capacity of bioweapons to inflict mass damage (Dando 2005: 36). Philip Coyle (2003: 11) comments that 'biological agents can certainly produce horrific deaths, but they make lousy weapons of mass destruction'. Biological agents are also extremely fragile, whereby they are highly susceptible to factors such as sunlight and adverse meteorological conditions. The fragility of these armaments makes them challenging to use to any significant degree. The extensive time that bioweapons need to take effect may also raise a question mark over their effectiveness. While a slow-acting weapon may be beneficial for some strategies, it may be entirely useless for more immediate forms of warfare. Furthermore, biological destruction can also be mitigated through access to protective clothing, decontamination, and preventative measures such as vaccines (Spiers 2010: 13). As such, there is no guarantee that an enemy cannot preclude the effects of biowarfare—regardless of what those effects might actually be. Consequently, the non-utility argument can still stand. The problem with that claim, however, is that this line of argument can, and has been, challenged in respect of the evidence outlined within the previous paragraph. At the very least, there is no clear proof that we can argue the point decisively either way.

Further on this contested issue of biological mass destruction, it should be noted that the same argument can also be turned into an opposing claim that these armaments do not possess utility. In this case, the assertion would not be that these weapons cannot cause destruction, but that the destruction created is too great. The argument is seen to reflect John Mueller's (2001) statement on the obsolesce of war, cited by Tannenwald with respect to her analysis. Mueller states that international actors are deterred from extreme forms of conflict having experienced its excess during the two World Wars of the twentieth century. Tannenwald (2007: 42–3) demonstrates, however, that Mueller's claim is not necessarily convincing. If actors were restrained by fears of extreme conflict, why have we not seen the tactical use of smaller nuclear weapons—and, by extension here, also smaller bioweapons? Bioweapons can be employed on a more limited scale than mass destruction (especially where this strategy involves non-contagious agents, such as anthrax) and also against non-human targets, for example crops. The restraint shown in not employing 'lesser' modes of violence would suggest that Mueller's argument is limited at best. Tannenwald's rebuttal is extended here to highlight that Mueller's claims can also be re-interpreted within the context of a taboo. Policymakers' antipathy towards extreme violence can be constructed as a normative dynamic—not least where extremely destructive tactics are considered unacceptable due to inherently moralistic concerns such as indiscriminate targeting. The issue of excessive warfare can be problematized beyond a practical calculation of effect to interrogate the issue of why policymakers are averse to that destruction, where Mueller is essentially talking about a deeply felt and emotive refusal to engage with mass killing as an intrinsic wrong. As Mueller (2001: 9) comments, that obsolescence connects to the way in which war 'has come to be regarded as repulsive, immoral, and uncivilized'.

Claims that biowarfare comprises a relatively cheap form of violence have also been contested (Ben Ouagrham-Gormley 2013: 474). This book demonstrates the point in Chapter 4, which analyses the US bioweapon programme's development during World War II. The multi-million-dollar project that, at its height, employed almost 4,000 people across numerous scientific sites in an industrial-scale programme is testament to the highly expensive, expansive, and extensive nature of biowarfare (Meselson 1999: 52). Disputing the cheapness of bioweapons does not, however, necessarily undermine the argument that bioweapons do possess utility. Tannenwald (2007: 39) notes the presence of extensive state programmes to conclude that they indicate a belief in weapons efficacy. Actor willingness to invest substantial levels of money and resources in maintaining significant arsenals would suggest that they do perceive utility; otherwise, why invest so much? Tannenwald's argument can be applied here to the development of costly biowarfare programmes. While the development of a major programme may not seem to comply with the terms of the taboo (a point discussed further in Chapter 4), it does raise significant concerns as to whether we can reduce the bioweapons issue to a simple lack of utility.

Admittedly, the presence of significant arsenals could be interpreted as a commitment to deterrence. Actors construct bioweapons not to employ them, but to deter. Yet the capacity for biological deterrence has also been disputed. The uncertainty concerning bioweapons' effect undermines both the possibility of a credible response and even the threat of use in the first place (Galamas 2008). And if bioweapons are used for deterrence, this argument again suggests that these armaments possess utility, both in the sense that they would be used initially and as a form of retaliation—a claim which further undermines the argument that avoidance is merely a product of weapon ineffectiveness. The pro-deterrence position would, in turn, raise further questions as to why first-use is still rejected. Tannenwald (2007: 30–8) develops the argument against deterrence in even more depth within her analysis of nuclear arms. She claims that the deterrence argument is insufficient in that: nuclear weapons states have not used these armaments even in cases where there was no threat of nuclear retaliation; nuclear arms have not deterred other types of attack by those who did not possess these weapons; non-nuclear weapons states have not been left powerless as a consequence of non-possession; and the vast majority of states do not own nuclear armaments and so apparently fail to value a deterrence capacity. Tannenwald's claims can again be applied here to bioweapons to challenge whether the deterrence argument fully explains the rejection of biowarfare on the same grounds.

The difficult nature of these alternative explanations for the rejection of bioweapons provides an interesting context for the wider study undertaken here. The real problem with these problems, however, is that these explanations do still apply. This book will show that certain actors do cite these arguments in order to justify their respective positions pertaining to biowarfare. Such arguments may never be shown to win out entirely over the taboo within a given scenario. The explanations, however, still comprise valid claims that are put forward by actors who often appear to genuinely believe in them or pursue them for strategic purposes. Indeed, Tannenwald does argue that the 'non-norms' explanations which she seeks to undermine (in order to identify the taboo as an explicative factor) still have relevance with respect to political decision-making. Her empirical research demonstrates that these explanations can comprise at least a contributing factor to nuclear warfare policy. Taking this point further, Tannenwald acknowledges that the taboo also fails to constitute an exclusive justification for actor behaviour in itself: 'The taboo is not the sole explanation for the non-use of nuclear weapons (non-use has occurred for other reasons), but it is an essential part of explaining this phenomenon' (Tannenwald 2007: 43). Consequently, Tannenwald (2007: 4) concludes that any analytic understanding of a taboo requires an investigation of both 'the mutual shaping of norms and interests'. The key question, therefore, becomes one of how these different explanations interact within the context of a given case. More specifically, to what extent did a taboo influence a specific outcome relative to other pressures that may also have been at play?

A concept of threshold

Price (1997: 122) discusses the contextuality of the chemical weapons taboo in relation to the concept of threshold. Threshold is the claim that a taboo operates whereby it 'raises the bar' against the use of pariah weapons or other prohibited behaviours. A taboo increases the costs of non-adherence to it. The costs involved may not necessarily preclude the impact of material interests and self-interested behaviour, that is, the point at which the threshold is crossed. The taboo, however, vastly restricts the chances that the norm will be transgressed. Taboos 'rarely render violations impossible but instead make them unlikely by raising the threshold of what counts as a legitimate exception to the rule'.

Consequently, Price also does not accept taboos as an absolute explanation for actor behaviour. Taboos, however, are still seen here to set the wider terms of both adherence and violation in demarcating the unacceptable. Price further argues that a taboo comprises a case in which alternative restraints would have been insufficient to prevent transgression in the absence of a taboo. A taboo constitutes a 'necessary condition' of upholding the norm in question and ensuring against contravention, even where this is then subject to other pressures (Price 1997: 110). Price considers taboos relatively strong in this respect. Threshold is portrayed here as an intrinsically extreme conceptualization in the sense that it constitutes a line which would only be crossed as an act of significant desperation. Price (1997: 121) talks of 'desperate thresholds' in order to reflect this point. Price—like Tannenwald—is not disputing that other matters are relevant. He maintains, however, that these other considerations cannot be explained in the absence of, and comprise only an additional explanatory factor to, a taboo.

The concept of threshold, however, risks setting up yet another straw person claim. Threshold can become a reductionist and binary tipping-point between adherence and non-adherence to a taboo—whereby non-adherence is equated to the realist position in the case of Tannenwald. This chapter, however, argues that such an assumption would comprise a problematic oversimplification of a taboo's relevance and function. Admittedly, Tannenwald does address the interplay between a taboo and what she identifies as the realist position. In doing so, she reiterates her claim that the taboo itself cannot be considered a sole explicative concerning actor behaviour. Tannenwald (2007: 4) then states: 'Norms often do not determine outcomes, they shape realms of possibility. They influence (increase or decrease) the probability of occurrence of certain courses of action.' Yet given Tannenwald's wider construction of realism as a 'non-norms' position, the threshold can still imply a case in which a taboo restrains up until a point at which other 'non-norms' factors can kick in. And even if this situation is not deemed to be applicable, there still remains a question as to how we interpret a 'realm of possibility' and apply this methodologically to the analysis of a taboo.

In response, this chapter asserts that our investigation of taboos needs to return to the model of taboo outlined in the previous chapter as a means of clarifying the approaches proposed by both Tannenwald and Price. A compliance-centric approach encourages the zero-sum thinking that undermines the concept of the threshold. Current thinking on taboos additionally holds our analysis back from developing a more specific notion of what 'shaping the realms of possibility' means by failing to fully establish the characteristics that define a taboo. Any analysis must be able to consider the more conditional aspects of a taboo in order to understand the complexities of its normative influence. Analysis must also then be able to contextualize that behaviour in respect of disgust, stigmatization, and fetishization as a means of identifying how those 'realms of possibility' function. Applying the new model of taboo here, therefore, uncovers the influence associated with a taboo, whereby these factors can then also be measured as an assessment of a taboo's relevance. Studies of a taboo do not need to make an unfeasible choice between the taboo and alternative explanations, where either is expected to apply or fail absolutely. When we engage fully with the intrinsic conditionality of a taboo, we can explore the entire scope of its influence and the ways in which a taboo shapes biowarfare policies.

Rethinking compliance; rethinking the threshold

The new model of taboo engenders methodological as well as conceptual implications for analysis. In particular, by more fully articulating the wider characteristics that define a taboo, the model underpins the claim (in line with Montgomery and Carpenter) that a taboo is not required to function as absolute in order to qualify its existence. Violation does not necessarily negate either the presence or influence of a taboo. A strict compliance-centric approach can, and does, deny a taboo on the basis of any example wherein actors do not appear to fully recognize or respond to normative constraint. Yet the new model maintains that contravention does not automatically repudiate. Consequently, there exist instances in which a taboo may not be identified as such on the grounds of that concept's definition where it would then be so under the terms of the new model—or, indeed, vice versa.

In the case of bioweapons, examples are discussed within this book that could be interpreted to contest the presence of a taboo in respect of biowarfare. For example, US development of a major biowarfare programme during World War II could potentially be cited to demonstrate that there was no effective taboo. Not dismissing the bioweapons taboo outright on this basis, however, permits the analysis here to consider that the taboo still influenced the decision-making process to a significant degree. The example cannot be fully comprehended without reference to the normative expectations which shaped both the programme and its outcomes—including the programme's ultimate renunciation. Such shaping would prove decisive to those outcomes, if not an absolute adherence to taboo principles. Consequently, the model opens up the debate on taboos simply by indicating that there may be more, and more influential, taboos than assumed. Taboos may be more pervasive within international political behaviour than some quarters have previously given credit for. By 'rehabilitating' taboo from the context of its own violation, IR analysis has greater scope to identify the presence of taboos, including the one pertaining to biowarfare.

The argument put forward here is not a claim that violation is always excusable in respect of the identification of a taboo. If a taboo is repeatedly contravened, this may question its status as such. If actors do not adhere to the expectations established by a given taboo, it would then be difficult to maintain that the norm continued to exist—at least in terms of what the taboo means in practice. Consequently, the rejection of a compliance-centric approach does not mean disregarding adherence entirely. Compliance remains a crucial aspect of measurement—in the same way as would apply to any norm. Cristina Bicchieri (2017: 50–1) highlights two necessary issues in respect to measuring a norm. First, analysis must prove a consensus that the norm exists. The identified consensus does not have to guarantee that actors will conform to the normative standard but must only establish that a social expectation is present. The second factor, however, does address compliance. Will actors obey the expectation? As such, compliance is still part of the analytic equation and an effective indicator that a taboo both exists and has sufficient strength to influence actor behaviour. The model proposed here, however, still challenges the necessary rejection of a taboo on the basis of compliance as an intrinsic aspect of the concept's definition. A key distinction exists between ascertaining a taboo's relevance and determining the presence of a taboo purely on the scope of compliance as a definitional standard.

Compliance then is not only an issue about the identification of a taboo, but also its political function. Under the terms of the new model, the analysis can now consider behaviours and outcomes that are not limited by a binary between pure obedience and pure violation, but which additionally relate to more conditional and secondary effects. This observation plays directly into Tannenwald's idea that taboos 'shape realms of possibility' even if they do not absolutely determine a specific outcome. By fully articulating the wider standards to which a taboo can be held (disgust, stigmatization, fetishization), analysts can speak of taboo-related behaviours that are not exclusively compliance issues (e.g. being disgusted, regardless of normative adherence) but which also indicate the presence of something 'more' than a non-taboo norm.

For example, the new approach can examine cases wherein the taboo has only partially influenced behaviour. Such cases would include a situation in which an actor may not fully or genuinely accept a taboo but is still influenced to the extent that others do or are believed to do so. As mentioned in the previous chapter, actors may feign compliance for strategic purposes such as upholding their reputation. Under the logic of a compliance-centric approach, the actor's own non-acceptance would be seen as a failure to adhere to a taboo. This failure then questions whether the taboo itself exists. Such logic, however, looks only at the individual, not the wider context—to see that the person at the centre of the example did not entirely accept the taboo. Yet, to the extent that the taboo is adhered to more widely, the same situation can still be interpreted as an indicator of normative strength. The actor both believes that others uphold the expectation and reinforces the norm via their conformity. Analysis, therefore, can still speak of an actor being 'affected by', even if this does not constitute personal 'adherence to', a taboo at the level of that individual. The subject still acts in accordance with, or responds to, the taboo in ways that acknowledge the norm as valid.

Consequently, it is argued here that strategic behaviours do not necessarily invalidate a taboo. To return to Cook's example of the women who would not eat with men, the explorer further observed that the women would eat with Cook and his male company as long as no one found out. Cook (2014: 91) concluded that 'whatever may be the reasons for this custom [the taboo], it certainly affects their outwards manners more than their Principle'. While the motivation and authenticity of compliance was questionable, therefore, the taboo remained intact and observed to most purposes. Moreover, Cook still recognized that the custom was a taboo. It may also be the case that even insincere compliance, if repeated, can become socialized. Friedrich Nietzsche (1910: 70) argues of the hypocrite that they can genuinely become what they feign to be: '[F]or instance, priests, who as young men are usually conscious or unconscious hypocrites, become at last natural and are then really without any affectation, just priests.' As such, this study maintains that disingenuity should not automatically lead to a taboo's negation. An example of strategic, fabricated, or even inconsistent behaviour may in fact comprise yet another indicator of a taboo.

A taboo can also be located in situations where the norm has only partially influenced an outcome. A case of this type would include a limited concession as opposed to all-out adherence; such as if an actor concedes on some aspects of a taboo to produce a compromise. Ultimately, a concession would technically only constitute a partial acceptance of taboo and so—to the extent that full compliance is demanded—a failure of that normative expectation, that is, the absence of a taboo. Abandoning the requirement of absolute compliance, however, means that less-than-perfect manifestations of a taboo can come to the fore. Furthermore, that by effectively highlighting 'new' ways in which the influence of a taboo can be identified, the model again supports a claim that taboos exert even more sway than certain aspects of IR have acknowledged. Failing to incorporate these means of influence into IR analysis ignores the full scope of what taboos do within the international political community.

The issue of a taboo's partial influence on outcomes relates to criticisms of Tannenwald for her selective focus on the non-use of nuclear weapons in respect to a taboo. Tannenwald effectively applies a compliance-centric approach to her analysis in that she only identifies a taboo concerning the first-use of nuclear armaments. She effectively concludes that the taboo does not extend further as actors are willing to possess nuclear armaments. In drawing this conclusion, Tannenwald has been challenged for overlooking the full picture of how actors understand and construct these weapons (Eden 2010: 837; Walker 2010: 869). It is argued that the nuclear taboo cannot be reduced to a question of weapons use, but must consider all aspects of actors' understanding of, and engagement with, nuclear warfare. Carol Atkinson (2010) said:

> I argue that acts of force, such as detonating a nuclear weapon, are only one aspect of how nuclear weapons are used. By restricting her meaning of 'use' to a material phenomenon, the detonation of a nuclear weapon, Tannenwald provides us

> with only a portion of a constructivist analysis of nuclear weapons policy by the United States. That is, in addition to thinking about how ideational mechanisms might lead to material outcomes, we should not forget that outcomes might also be ideational.
>
> **(Atkinson 2010: 839–40)**

Atkinson argues that these wider factors relate to issues such as possession and associated perceptions of power, whereby that power is expressed in ways other than the physical employment of an armament. For example, Atkinson asserts that possession can confer a capacity for coercion or a sense of prestige that is not available to the non-nuclear weapons state. The same points are applied to biowarfare in this study. In doing so, removing the focus on compliance permits a level of conditionality which can then be applied to examine whether and how a taboo can still function in respect of these other factors. In particular, this type of analysis can address the issue of fetishization and the way in which a threat is prioritized as a special concern, which does not require an example of use in order to be upheld.

Conditionality

Rethinking compliance also opens the analysis up to the conditionality of a taboo in respect of individual actors. The argument here is that any taboo is not, and should not be considered, universal. Different actors will adhere to a taboo to different extents for different reasons, including moral preference. While Thomas Dolan (2013: 41) discusses the restraining authority of taboos on weapons use, he further highlights that 'no person is perfectly socialized, and therefore even members of a single organization can differ in casual beliefs, interpretations of events, and fundamental values'. Socialization can be viewed as a spectrum on which some actors are more influenced than others. This statement is not necessarily controversial. We already know that some people murder, say, and others do not—whereby standards of behaviour are not uniform. The identified conditionality can also be comprehended within the context of groups, organizations, and institutions. Kathryn Sikkink (2013: 162) argues that powerful groups of actors can influence norms depending on, and in accordance with, their own internalization of meaning—which may not be the same as other groups. Collective units can express interests and beliefs that drive a different or no adoption of a taboo relative to others based on localized and bureaucratically driven notions of what is deemed permissible. Groups may also be defined by exclusive cultures and bureaucratic interests which shape, and even negate, perceptions of a taboo. Normative conditionality may also relate to group survival. A certain group may have a vested interest in upholding a taboo, or not, according to their needs and the preservation of their institution.

Admittedly, this issue of conditionality then raises important questions around who holds the power to influence a taboo, that is, who possesses the sway to exert or deny one. If an actor does or does not adhere to a taboo, it may not matter if they are an individual citizen. If the actor is the US president, however, this becomes quite another matter. The principle remains, however, that contextuality must be factored into any analysis of a taboo as opposed to a case where universal adherence is taken as an expectation. The approach then creates the basis of a more comprehensive discussion of what restrains international actors.

Violation and normative reinforcement

Violation can even reinforce a taboo. Violation is frequently associated with normative irrelevance and erosion. The transgressive act ostensibly signals that actors are unconstrained by, and do not fully acknowledge, the norm in question. Yet norm-breaking can have more constructive effects than invalidation—although this is not to state that invalidation cannot also occur (Deitelhoff and Zimmermann 2020). Three key points relate to this, drawn from the existing literature on normative non-compliance. First, violation is only disruptive if it generates a subsequent cascade of non-adherence to a norm (Thomas 2000: 132; Glennon 2005; Panke and Petersohn 2012: 721); what Gregory Raymond (1997: 218) calls 'a rising tide of violations'. A single transgression (especially a minor one), or the containment of transgression to one actor (especially a pariah, e.g. terrorists), would not necessarily generate norm erosion. As Elizabeth Keir and Jonathan Mercer (1996: 91) state: 'Just as it requires followers to set a precedent, it requires followers to kill a convention. This explains why violating a convention does not necessarily kill it.' Moreover, normative regimes may be supported by 'norm clusters' that offer wider support to the transgressed rule and so preclude a cascade (Lantis and Wunderlich 2018). Norms are 'insulated from contestatory challenges' by the resilience of multiple other norms to which they relate as part of a wider regime (Lantis and Wunderlich 2018: 570).

Second, a violation may not disrupt a norm if the act is in some way justified. Even the mere offer of a justification—regardless of whether this constitutes a 'good' excuse—denotes a continued acknowledgement of the norm whereby contravention is effectively defended as an aberration (Kratochwil 1995: 63; S. Smith 2001: 209; Cortell and Davis 2005: 8; Sandholtz 2008: 108–9). The 'apology' recognizes that the broken taboo represents a standard that should have been upheld, even if it was not in a given instance. This observation is especially relevant if the violation was unwilling or in some other way conditional. As Bela Szabados (1979: 200) comments on the issue of hypocrisy and moral belief: 'I believe that someone can sincerely hold a moral belief and yet not act in accordance with it on some, perhaps critical occasions.' Critically, the justification does not have to come from the violating actor. Others may rationalize the transgressive actions as a means of upholding a given norm, for example by negating the deviant actor as a pariah.

Third, normative perception may be reinforced if a violation is subsequently chastised or subject to punishment. Brad Roberts (1992: 104) said that 'norms matter in international politics—not because they constrain the choices of the most malevolent of men but because they create the basis for consensus about responses to action inconsistent with those norms.' A response that challenges the transgression upholds the norm as valid, significant, and creates a disincentive to further transgression (Legro 1997: 35; Panke 2010; Panke and Petersohn 2012). The response can also promote normative adaptation to challenge (Smetana and O'Mahoney 2022). Furthermore, actors are reminded through their reaction of the need to commit to a norm and also reassert their own commitment to that norm through its enforcement. Overall then, violation is not only not an automatic signal of the non-existence of taboo. It may be part of the dynamics which actually cement the prohibition.

Contextualization

Compliance must also be rethought in respect of its contextualization, that is to say, normative adherence is underpinned by the more expansive characteristics of disgust, stigmatization, and fetishization. Contrary to Peter Gizewski's (1996: 399) claim that taboos 'are more comprehensible when the term [taboo] is applied in a "figurative" or loose sense', this book asserts that an even more detailed definition is required. In applying this more specific approach, analysis on taboos would not only search for evidence of compliance, but also actor perceptions of the taboo subject. Is the actor disgusted, including in explicitly moral terms? Do they demonstrate evidence of stigmatizing behaviour and expression, such as acknowledging stigmatization processes, for example formal codification? Do they fetishize the subject, for example demonstrate a prioritization of the taboo? The last line of questioning is especially important given that the issue of fetishization has been overlooked in respect to IR analysis of taboos. The sacred and prioritized elements of a taboo are critical in comprehending why the concept is distinct from non-taboo norms and to avoid the reduction of a taboo to the latter.

Answers to the above questions can be addressed in the presence of normative violation. In particular, the analysis can ask whether the terms of violation reinforce, and are expressed in terms of, the broken taboo. For example, does the transgressive actor display remorse at their own taboo-breaking and accept the associated disgust and immorality of their actions—where this response demonstrates an acknowledgement that the taboo exists and is relevant (if not followed in a specific example of behaviour)? Furthermore, analysis would also search for partial evidence of taboo. Are there instances where taboo sentiment still influenced a given decision, such as through the negotiation of compromise or concession? While compliance itself remains an analytic consideration, the alternative approach proposed here suggests a type of study that also engages fully with other characteristics. Indeed, it is argued in this book that these characteristics must be explored if an analysis of a taboo is to be complete.

Taboo talk as a method

Measuring a taboo entails four key and interrelated lines of investigation. The four lines establish: (1) is there evidence of a taboo as a distinct type of normative prohibition; (2) does that taboo hold in some way (regardless of whether this comprises an intrinsic condition of point 1 as a reference to compliance); (3) what does it mean for a taboo to hold; and (4) did other factors influence an outcome and to what extent?

Tannenwald demonstrates that the answers to these questions lie in what actors say; more specifically, how actors both reason and justify their position. There are two core elements to this methodological approach. First, the method involves a process-tracing exercise which identifies evidence of decision-making with respect to a given situation or situations. This book achieves that objective through an archival study of selected case studies. Second, Tannenwald (2007: 51–4) cites actor talk as the key aspect in measuring the presence and influence of a taboo—which she terms 'taboo talk'. Taboo talk is interpreted as explicit or implied references to a taboo. The references are presented by the actor as informing their judgement in respect of their associated behaviour and communicating opinion in the terms and phrasing of a taboo. If a taboo exists and is relevant, analysis would see actors evaluating a given situation in relation to the constraints of that taboo. Critically, however, Tannenwald also cautions that in looking for talk we must be careful of silence. What people do not say may be equally as telling as what they do. The point is especially pertinent to the taboos surrounding pariah weapons in that silence can relate to wider issues of secrecy. Secrecy (as already shown in the introduction to this book) can be indicative of the presence of a taboo and not its absence.

This study adopts Tannenwald's method as a form of qualitative discourse analysis. It is maintained that investigation can 'find' a taboo within the ways that actors express themselves. The approach applied here, however, still reflects the model of taboo proposed within this book. In examining actors' available written and spoken statements for references to the taboo, analysis must specifically seek to uncover evidence of disgust (as well as related taboo concepts such as immorality), stigmatization (which includes the identification of codified documents such as international agreements), and fetishization (talk that ascribes excess and sacred value to the subject of a taboo). The three strands of taboo, therefore, provide the core structure of the data analysis.

A discourse-based methodology must be applied in ways that recognize the strategic potential of rhetoric. Actors may say things for reasons other than genuine personal belief, for example to uphold their reputation or to fulfil other strategic ambitions. It has already been demonstrated elsewhere that taboos are not simply a prohibitive pressure on actors but are also a resource for strategic speech (Bentley 2016). Tannenwald touches on this strategic element of rhetoric. In doing so, she advises that analysis should resist any temptation to reduce 'taboo talk' to 'cheap talk'. It is not the case that we can dismiss what politicians say because we do not think that they mean it. Taboos, therefore, must be examined within the context of their deliberate, persuasive, and strategic employment relating to both domestic and international political rhetoric.

Strategic rhetoric is discussed in this study within the context of strategic narratives. The theory of strategic narratives draws on the 'discursive turn' within IR represented by scholars such as Janice Bially Mattern (2005) and Charlotte Epstein (2008). The concept is defined as:

> [A] means for political actors to construct a shared meaning of the past, present, and future of international politics to shape the behaviour of domestic and international actors. Strategic narratives provide a tool for political actors to extend their influence, manage expectations, and change the discursive environment in which

> they operate. ... The point of strategic narratives is to influence the behaviour of others.
>
> **(Miskimmon, O'Loughlin, and Roselle 2013: 2)**

Strategic narratives are the words, speech, and discursive actions that allow actors to present politics in specific ways. They are the 'compelling storylines' by which actors utilize rhetoric to craft and enact perceptions of a political situation or issue in order to shape opinion in line with their own ambitions (Freedman 2006: 22; see also Hajer 1997: 56; O'Tuathail 2002: 617; Dimitriu 2011: 195). Actors use narratives to create a description of the world—that is, an understanding of what politics looks like and entails—whereby they 'force' their audience into adopting certain interpretations and pathways of logic that complement the aims of the constructing actor.

Significantly, the reference to the strategic does not mean that a given actor does not genuinely believe the ideas that comprise a narrative. While the theory of strategic narratives identifies a calculated element to the application of those ideas within certain forms of rhetorical presentation, the theory does not assume that the commitment to those ideas is a fabrication. Even if the rhetoric under scrutiny represents a partial or feigned commitment to a taboo, however, strategic rhetoric remains relevant to proving the relevance of that taboo—to reiterate the point already made above. The strategic utilization of X necessarily entails that an actor believes that X exists, even if they do not personally accept X. Moreover, the actor demonstrates that they further believe that X is sufficiently accepted by others to the point that X has weight and can be utilized effectively for strategic purposes. If no one else accepted X, then there would be little point in developing a narrative around it. Consequently, analysis can both accept that actors state things that they do not intrinsically or entirely accept because they think they ought or are strategically motivated to and that (1) disingenuity does not negate the presence of 'real' commitment within other utterances of political speech and (2) even if speech is contrived in some way, that the strategic construction still reveals something important about why speech is phrased in the given terms, that is, the taboo. Strategic narrative, therefore, is still evidence of a taboo's relevance.

A new means of measurement

This chapter has outlined the core principles for measuring a taboo, with special reference to IR. In particular, the chapter has critically reassessed the approach adopted by Tannenwald within the context of the new model of a taboo proposed in this book. Adopting the model has both conceptual and methodological implications for the measurement of taboos. By more fully articulating the characteristics of taboo and rejecting an exclusive focus on compliance as the basis of a taboo's identification, the model permits an expansion of the debate on taboos to consider a significantly wider range of examples of normative bearing. These other examples are ones which may previously have been excluded as irrelevant or as evidence of a taboo's non-applicability under previous interpretations. Under the terms of the new model, studies of taboos can now engage with a more expansive selection of cases which are related to taboo behaviours. This evidence is found within the speech and written utterances of the actors who express a taboo, that is, taboo talk. These expressions specifically relate to references of disgust, stigmatization, and fetishization as the basis for locating a taboo. What actors say may not always be genuine in the sense of absolute compliance. Yet these statements still reveal much about what taboo is, how it functions, and how these normative prohibitions restrain us in so many aspects of our lives. The following empirical chapters now apply this understanding to the issue of biowarfare and the taboo that surrounds this threat.

3
Codification
Ancient Warfare to the Geneva Protocol

Codification is an extremely important factor in respect to a taboo. Price (1997: 18) argues, for example, that the chemical taboo cannot be realistically identified until its codification and politicization through official diplomatic and legal prohibitions. Jefferson (2009: 54) describes Price's argument as a case in which codification comprises the very '*cause*' [emphasis in original] of the taboo itself. Price's claim could be seen to preclude much of the chemical taboo's early history. His statement would overlook the disgust, informal stigmatization, and fetishization inherent to the social construction of the chemical taboo that exists prior to, and irrespective of, the norm's formalization within international agreements. Yet the recognition of a taboo in codified form can still be identified as a critical marker in a taboo's normative evolution. As Link and Phelan (2001: 375) comment, 'it takes power to stigmatize'. It can then be claimed in response to their statement that codification provides both the institutional and political means for the exercise of that power. Codification gives weight to the enactment of stigmatization (as well as to disgust and fetishization) as an established and workable feature of socio-political life. Codification further cements a taboo by providing normative clarity. Amy Smithson (2004: 166) demonstrates that global norms are frequently difficult to quantify. She then argues that treaties are more precise in their expression of normative meaning and so give greater cogency to the norm in question. In doing so, norms are subsequently less likely to be disputed and questioned. Smithson said: 'Thus, treaties possess a tangible quality that norms lack.' Consequently, codification does comprise a key aspect of a taboo's presence and robustness.

Price's link between codification and the chemical taboo is expanded here in relation to bioweapons. The link to chemical weapons, however, is not simply a point of comparison or a normative model that can be applied to other examples. Price's comments on chemical arms are of special import in that the early codification of bioweapons was intrinsically connected to

The Biological Weapons Taboo. Michelle Bentley, Oxford University Press. © Michelle Bentley (2023).
DOI: 10.1093/oso/9780198892151.003.0004

perceptions of chemical warfare. International agreements prior to the 1925 Geneva Protocol typically spoke of 'poison' warfare, which comprised a reference to both biological and chemical violence. Even the Protocol itself simultaneously bans both forms of aggression. The Protocol is also vital to this analysis, however, in that the agreement makes a more explicit reference to bioweapons than was seen in the documents before it. The Protocol recognized biowarfare as a threat in its own right, even where that threat was considered in tandem with chemical armaments.

The early conflation of biological and chemical violence is unsurprising. It was only in the late 1880s that the 'germ theory' of Louis Pasteur and Robert Koch developed. The new theory advanced a more sophisticated understanding of disease and the ways in which biology was distinct from chemical processes. This scientific development also underpinned new ideas concerning the weaponization of biological agents and the technological means to produce such weaponry as different from those used in respect of chemical warfare (Wheelis 1999a: 8–9; McCauley and Payne 2010: 10; Barras and Greub 2014: 499). The early reference to poison was, therefore, not a deliberate and constructed conflation that analysis should read some deeper political meaning into. Biological and chemical aggression were viewed as part of the same threat in that both armaments contaminated the body and where there was no core scientific distinction to be drawn between them at that time. As scientific understanding separated the biological and the chemical, however, so arms control would eventually follow.

Consequently, this book still maintains its claim that the analysis of the bioweapons taboo should consider those armaments on fundamentally unique terms. Yet the conflation of chemical weapons with the biological is also recognized as an intrinsic aspect of the taboo's earlier history and a crucial issue in the normative rejection of bioweapons. This chapter outlines how this perception was expressed within the evolving codification of the bioweapons taboo and explores some key questions in doing so. How were bioweapons included within the concept of poison? How did codification engrain the taboo sentiments towards biowarfare outlined in Chapter 1? The chapter also looks specifically at the Geneva Protocol. Why was this agreement such a significant aspect of the taboo, not least where the text explicitly referenced bioweapons in a way not expressed within previous international agreements? How did the Protocol formalize the rejection of biowarfare? How did the Protocol grant customary status to the taboo?

Ancient cultures and early codification

The codified rejection of warfare that contaminates the body dates back to the Hindu *Laws of Manu* c. 200 BCE (Zanders 2003: 392). The document states: 'Fighting in battle, he should not kill his enemies with weapons that are concealed, barbed, *or smeared with poison* or whose points blaze with fire' [emphasis added] (*Laws*, 1991: 79). Jefferson (2009: 33) partially questions the extent to which the *Laws* ruled out the use of poison tactics. She highlights that elsewhere the code endorses the contamination of an enemy's environment during a siege: 'When he has shut up his foe (in a town), let him sit encamped, harass his kingdom, and continually *spoil his grass, food, fuel, and water*' [quotation and emphasis in Jefferson]. The *Laws*, however, can still be interpreted as a formalized rejection of poison as (at least some forms of) weaponry. While the document did not specify precisely what poison was, it can be inferred from the types of employment referenced that the term could incorporate both biological and chemical forms of violence.

Ancient Greek culture also rejected poisonous armaments as 'abhorrent' (Hersh 1968a: 3; see also Carus 2017a: 4). In the same vein as the *Laws*, poison arrows in particular were stigmatized as immoral in that they were seen to cause disproportionate, psychologically distressing, and especially barbarous suffering (a view replicated within the Islamic tradition, see Harbour 1995: 76). The Ancient Greeks interpreted the poison arrow as inflicting a 'double death': the death of the person and the death of their honour in dying by such means (Mayor 2009: 621).

The repulsion towards this mode of violence is repeatedly expressed within the Ancient Greek myths. Analysis can look to the tales of Heracles as an example. Heracles acquires his infamous poison arrows after killing the Lernaean Hydra as part of his Twelve Labours (Apollodorus 2008: 3i–5h).[1] Heracles slays the serpent-like monster and then dips his arrows into her vile venom to make them poisonous.[2] The arrows are now a highly potent weapon—yet they lead to the most wretched of consequences. For example,

[1] Olympian matriarch Hera caused Heracles to experience an episode of insanity in which he accidentally kills his wife and children. As Natalie Haynes (2020: 120) points out, 'this part of his story was wisely omitted from the Disney animated film'. Heracles was sentenced to complete Ten Labours set by King Eurystheus in penance for the killings. Eurytheus disputed Heracles' efforts on two of the tasks, including the Hydra's defeat. The king claimed that the Labour should be disallowed as Heracles' nephew Iolaus assisted him in killing the monster. Eurytheus then set two further tasks to make up for the shortfall: stealing a golden apple from the garden of Hesperides and capturing Cerberus, the three-headed dog that guarded the gates of the Underworld.

[2] Venom is a toxin—a chemical substance that is biologically produced. The issue as to whether a toxin should be classified as a chemical or a biological weapon is subject to dispute (as will be explored further in Chapter 6).

the arrows cause the death of the revered centaur and Heracles' friend, Chiron. Centaurs gate-crash a dinner that Heracles attends during his fourth Labour. The hero retaliates by attacking them with his poison arrows (Apollodorus 2008: 5). The interlopers take refuge from the strike with Chiron, who is accidentally hit by an arrow in the melee.[3] Chiron's death is a tragic scene. The beloved centaur writhes in agony and begs the gods on Olympus to be released of his immortality as he cannot endure his suffering and wishes to die. The excruciating pain—to the extent that Chiron prefers death—is a fetishized testament to the sheer levels of torment associated with poison weapons.

In a further myth dramatized by Sophocles, Heracles and his wife Deianira employ the centaur Nessus to transport them across a river (Sophocles 2008: 555–87). Nessus sexually assaults Deianira as he carries her. Heracles fires a poison arrow at Nessus in retaliation. As he dies, Nessus tricks Deianira into taking some of his blood which is now contaminated with the '[m]urky venom' of the Hydra (Sophocles 2008: 574). Nessus tells Deianira that the blood is a love potion. If Deianira ever needs to make Heracles desire her, she should use the blood and his love will be renewed. Later and after years away at war, Heracles returns home with a young woman named Iole—albeit not consensually on her part (Sophocles 2008: 381). Deianira sends Heracles a tunic covered with the centaur's blood in a desperate attempt to win back her husband's affections. Heracles puts the tunic on and experiences the most tremendous and gut-wrenching agony from the '[m]alignant viper's [Hydra] poison' (Sophocles 2008: 771). He endures a truly horrendous death. In Ovid's (2014: 7.229–31) version of the tale, Heracles' suffering is so great that he builds his own funeral pyre to kill himself upon. Like Chiron, death is preferable to the excruciating pain.

Heracles' demise expresses contemporary perceptions of poisoning as dishonourable, uncivilized, and intrinsically feminine. After so many heroic acts and glorious military victories, Heracles does not die bravely like a solider. No monster, god, or army—but instead 'a feeble Woman, without a drop of masculine blood, Alone and swordless, managed to bring me down' laments the hero (Sophocles 2008: 1061–3). Moreover, Heracles' excessive physical suffering from the poison causes him to perish while 'crying away like a little girl' (Sophocles 2008: 1072). Sophocles employs the stigmatized labelling of Heracles as a female child in order to articulate how ignoble this means of death is. Heracles' own disgust at the nature of his death is compounded

[3] Ovid (2004: 397–409) outlines an alternative version of Chiron's death. Ovid says that the centaur was in the process of examining one of Heracles' poison arrows when he accidentally dropped it onto his foot. Chiron dies as a consequence of the wound.

further by the fact that Deianira then kills herself with a sword. Ancient Greek women typically committed suicide with poison. The Chorus express their shock that a 'weak' woman could select such a masculine method of self-destruction and violence (Sophocles 2008: 898). Consequently, Deianira achieves the proper death that Heracles should have received. Her actions reinforce just how emasculating and undermining it was to die by poison in the ancient mind.

Biowarfare was alleged to have occurred in the ancient era.[4] Athens accused Sparta of contaminating their water supply in 430 BCE following an epidemic in the city during the Peloponnesian War. The outbreak is typically assumed to have involved a case of plague. The exact disease, however, remains unknown and may have instead been (amongst other suggestions) Ebola (Scarrow 1988), tularaemia (Wylie and Stubbs 1983), or typhoid (Papagrigorakis et al. 2006). One third of the Athenian population perished in the epidemic. The incident is captured in harrowing detail in Thucydides' *The History of the Peloponnesian War*. Thucydides (2004) comments that the epidemic was 'almost too grievous for human nature to endure':

> Such was the nature of the calamity, and heavily did it weigh on the Athenians; death raging within the city and devastation without. Among other things which they remembered in their distress was, very naturally, the following verse which the old men said had long ago been uttered: A Dorian war shall come and with it death.
>
> **(Thucydides 2004)**

Thucydides notes that the Spartans had been present in the area for a few days prior to the disease breaking out, hinting at the coincidence. Yet he also says of the allegations that he would 'leave to other writers, whether lay or professional' to determine the accusations (Thucydides (2004). Whether the Athens epidemic comprised a calculated act of biowarfare or not, the nature of the allegation and the description of effect still demonstrate the sheer fear, disgust, and repugnance associated with bioweapons at this time and the identification of biowarfare as an illegitimate means of conflict. Biowarfare was an aberration; a tactic that was associated with barbarians who would not hold to the principles of a civilized means of combat.

The perception of poisonous warfare as immoral continued through into the Roman era. The *Lex Cornelia de sicariis (et veneficis)* [Cornellian law of

[4] The reference to allegation (as opposed to a proven violation) is very important. Carus (2017a: 4) is extremely sceptical of claims that biowarfare was employed in respect of ancient warfare. He states that there is insufficient evidence to prove that the alleged incidents ever happened.

assassins (and poisoners)] was brought into force in 81 BCE (Fellmeth and Horwitz 2009: 167). The law condemned both the poisoner and any person who created or provided the means of poisoning. The punishment for these crimes was severe and could even involve being thrown to the wild beasts (Kaufman 1932: 166). Similarly, Cicero's (2014) *De Officiis* rejected forms of poison as a weapon. The philosopher stated that 'the poisoned cup' comprised the method of a person whom Cicero labels as 'not strictly upright' and who had rejected 'the moral right'. Valerius Maximus (1726: 574) also drew a line between modes of civilized and uncivilized aggression in respect to poison. He expressed the now famous words, '*armis bella, non venenis, gerere debere* [war is waged with weapons, not poison]'.[5] Interestingly, the concept of deliberate poisoning was often connected to notions of magic at this time. *Venenum* can mean both 'poison' and 'magic potion' (Pharr 1932: 269; Rives 2003: 320). Despite this supernatural element—indeed, in part because of it—the taboo against poisonous violence was now cemented within codified and legal proceedings. Christianity further introduced complementary ideas concerning moral conduct within conflict and Just War. Religious attitudes entrenched ethical beliefs pertaining to the immorality and inappropriateness of unnecessary suffering. This rejection of certain forms of harm as disproportionate would later appear in subsequent codified prohibitions as a means of distinguishing biological arms as intrinsically immoral and as necessitating special proscriptions on that basis.

Yet the allegations of biowarfare continued. The Carthaginian General, Hannibal, supposedly threw earthenware jars filled with venomous snakes onto the ships of his Roman-backed Pergamum enemy during battle in 190 BCE (Dawson 1967: 124). Accusations were also made against the Romans themselves and the later Eastern empire. It was said that the Holy Roman Emperor, Frederick Barbarossa, used the bodies of deceased soldiers to contaminate the wells of the Italian city of Tortona in 1155 (Kroll 2006: 21). In 1171 during the Byzantine–Venetian conflict, the Venetians claimed that they were the victims of a calculated plague attack against the water supply to their naval war fleet (Grygiel 2006: 69). The most infamous of these historical accounts, however, is the 1346 Mongol attack on the city of Caffa, in what is now Crimea. The invading forces are reported to have catapulted corpses that were contaminated with the plague into Caffa in order to break a

[5] Valerius Maximus' full quote is: 'Ea res cum ad senatum esset delata, missis legatis Pyrrhum monuit ut adversus huius generis insidias cautius se gereret, memor urbem a filio Martis conditam *armis bella, non venenis gerere debere*' [emphasis added]. Consequently, the reference is specific to societies that are 'founded by the son of Mars'. The statement expresses an idea that the revulsion towards contaminating weapons was associated with an explicitly civilized society.

siege. The example is well-known for the subsequent claim that the outbreak ultimately led to the pandemic Black Death of the Middle Ages in which up to 60 per cent of the global population died (Benedictow 2004: 382).[6] The attack on Caffa is recounted in the contemporary writings of the Italian notary, Gabriele de' Mussi. De' Mussi (in Di Cicco 2014: 39) said that there were 'mountains of dead men catapulted into the city as rain'. De' Mussi (in Deaux 1969) also discussed the subsequent spread of the disease from Caffa as people fled the city following the attack:

> Woe to us for we cast at them the darts of death! Whilst we spoke to them, whilst they embraced us and kissed us, we scattered the poison from our lips. ... Oh death! cruel, bitter, impious death! which thus breaks the bonds of affection and divides father and mother, brother and sister, son and wife. Lamenting our misery, we feared to fly, yet we dared not remain.
>
> **(De' Mussi in Deaux 1969: 76)**

Blankets

Ancient ideas that biowarfare was immoral influenced later thinkers on the ethics of war. Alberico Gentili (1933: 40a, 156) called poison warfare 'forbidden' and 'impious' in 1612. He further asserted that these tactics were the work of uncivilized actors. Poison was used by 'infidel barbarians' and that it 'is their way to use unfair wiles, to fight with poison, to wage a merciless war ... which has been abolished in all Christian warfare' (Gentili 1933: 402). Hugo Grotius (2005: 1290) similarly condemned the use of contaminating violence in his 1625 *The Rights of War and Peace*. Grotius commented that the Law of Nature might not seem to discriminate between whether a person is killed 'by the Sword or by Poison'. If that person died either way, should the means by which they are killed concern us? To return to the quote by Walt (2013) in Chapter 1: 'Dead is dead—no matter how it is done.' Grotius clarifies this stance, however, by referring not to the Law of Nature but to the Law of Nations. He states that by the social and legal standards of nations, 'if not of all, yet of the more civilized, allows not the taking the Life of an Enemy, by Poison' (Grotius 2005: 1290).

[6] Claims that the incident at Caffa caused the wider Black Death pandemic are disputed. Carus (2017a: 6) argues that it is unclear whether the corpses used in the strike could have carried the fleas that spread plague as the insects require a living host. It is further questioned whether one outbreak alone was sufficient to trigger such an extensive global event (Ditrich 2017).

The ideas expressed by Gentili and Grotius were effectively codified 50 years later within the 1675 Strasbourg Agreement. The bilateral deal between the French and the Holy Roman Empire prohibited the employment of poison bullets within the context of warfare. Strasbourg is identified as the first arms control measure in relation to chemical weapons (Zanders 2003: 394). The agreement, however, did not specify that the concept of the poison bullet to which the restriction applied was exclusive to chemical aggression. The terms of the agreement did not preclude the additional inclusion of armaments contaminated by biological means. Consequently, the document can still be seen to have contributed to the taboo against poisonous warfare more generally and further engrained the normative opposition to the utilization of disease as a weapon specifically.

Allegations still persisted that the taboo was being broken in respect to biowarfare. Twenty-five years prior to the Strasbourg Agreement, Polish artillery general Jan Kazimierz Siemienowicz reportedly ordered projectiles to be filled with saliva from rabid dogs and fired against the enemy (Dembek 2006: 18). French historian Pierre-Francois Xavier de Charlevoix claimed that in the early 1700s an indigenous group of Iroquois had contaminated a stream which supplied English military camps by throwing animal pelts into the water during the Queen Anne's War (Wheelis 1999a: 27) The attack supposedly killed over a thousand men and instigated a British retreat. Accusations of water contamination were also rife throughout the American Civil War. For example, archival accounts suggest that Confederate troops infected the local water supply when they were forced out of Vicksburg, Mississippi, in 1863 (Williamson 2012: 179). The soldiers supposedly sought to deny fresh water to the Union troops, who were then approaching, by leaving the rotting corpses of dead livestock in freshwater ponds. Union Colonel, James Slack, of the 47th Regiment Indiana Volunteers wrote in a letter to his wife: 'The rebels are burning the corn in the country to prevent us using it, and are killing hogs, cattle, and sheep, and throwing them into the water along the road to poison it so we cannot use it' (in Williamson 2012: 179).

Also infamous within the history books is a case whereby the British military purportedly infected native tribes in America with smallpox during the Pontiac Rebellion in 1763. William Trent was a partner in an Indian trading company called Levy, Trent & Company and had been commissioned as a captain at Fort Pitt in Pennsylvania. His diary from the time reveals the supposed plot. Trent (1924: 400) recalled that there had been a local outbreak of smallpox which had spread to the Fort. At the same time, Trent recounted that a group of native Americans had come to the Fort to discuss the ongoing war. He said: 'Out of regard to them, we gave them two Blankets and a Handkerchief out of the Small Pox Hospital. I hope it will have the desired effect.' Trent's reference to the 'desired effect' is interpreted as a deliberate attempt to infect the native Americans.

Carus (2017a: 7–8) advises caution in making an allegation of biowarfare in relation to Fort Pitt. In particular, he argues that it is problematic to accuse Sir Jeffrey Amherst, then Commander of the Southern Department of the British Army in North America, of carrying out the subterfuge.[7] Yet there is evidence that Amherst at least considered the idea of using smallpox in an offensive manner against the native Americans. Colonel Henry Bouquet had sent Amherst (Bouquet's commanding officer) a letter dated the day prior to the blankets allegedly being distributed in which Bouquet advised Amherst of the smallpox outbreak. Amherst replied:

> I cannot but wish, that Whenever we have any of the Savages in our Power, who have in so Treacherous a way Committed any Barbarities on our People, a Quick Retaliation may be made, without the least Exception or hesitation.
>
> **(Amherst in Knollenberg 1954: 490)**

Amherst (in Knollenberg 1954: 492) further speculated: 'Could it not be contrived to send the small pox among the disaffected tribes of Indians? We must on this occasion use every stratagem in our power to reduce them.' The letter certainly suggests that Amherst advocated the ideas inherent to the alleged plot. In what is possibly a further example of this tactic, Carus (2017a: 9–11) also notes that at least one of the officers at Fort Pitt was later stationed at a new British colony established in Australia. A devastating smallpox outbreak occurred there among the indigenous Aboriginal population in 1789—just a year after the colony was opened. Carus highlights that the epidemic was extremely convenient for the British who were attempting to subdue the local population at the time. The implication is that the tactic of biowarfare was learned at Fort Pitt and then employed elsewhere.

The rules of war

The type of bioviolence tactics supposedly used at Fort Pitt would later be officially condemned in the US. On 24 April 1863, a set of guidelines for conduct in warfare by the US Army was signed into force by President Abraham Lincoln. *General Orders No. 100* was also known as the Lieber Code after its author—the lauded ethicist Francis Lieber. Lieber was a former solider and a

[7] For a further account of the controversies surrounding the allegations see Philip Ranlet (2000).

veteran of Waterloo. He had a long-standing interest in conflict ethics (Nys 1911: 355) and has since been described as having 'originated the idea of codifying the laws of war' (Gilman 1901: 274). President of the American Society of International Law, Elihu Root (1913: 466), once said of Lieber: 'If our Society, at once national and international, were about to choose a patron saint, and the roll were to be called, my voice for one would answer "Francis Lieber".' Lieber (in Davis 1907: 20) commented of his motivation in constructing his Code that he was driven by the taboo ideas of morality and civilized modes of fighting: 'Usage, history, reason, and conscientiousness, a sincere love of truth, justice, and civilization have been my guides.' On the issue of poisonous warfare, Article 16 of the Lieber Code (*General Orders No. 100*, 1863) refused to 'admit of the use of poison in any way'. Article 70 further stated:

> The use of poison in any manner, be it to poison wells, or food, or arms, is wholly excluded from modern warfare. He that uses it *puts himself out of the pale of the law and usages of war* [emphasis added].
>
> (*General Orders No. 100*, 1863)

Lieber again did not explicitly differentiate between biological and chemical aggression and his restrictions against poison could be applied to either.

Significantly, the terms of the Code did still engage with ideas of military expediency more widely and did not prohibit all forms of violence in warfare. For example, the Code's provisions recommended that an army commander should be willing to adopt a hard-line and even creative approach to how they defeated their enemy—including the implementation of more unconventional tactics such as starvation methods. Consequently, Lieber is described as 'no "mawkish sentimentalist"' (Witt 2014: 160). The deeper context outlined here is relevant to this study whereby the observation demonstrates the extent to which poison warfare stood out as an especially horrific and disreputable form of conflict. In a situation wherein Lieber was prepared to consider other morally questionable forms of conflict as permissible, it was poison which was considered a step too far.

Lieber's principles were adopted internationally within the 1868 St Petersburg Declaration (Carnahan 1998: 213). The agreement was 'vague' (Neinast 1964: 25) and did not make any explicit reference to poisonous warfare. The declaration did, however, seek to reinforce the ethical ideal of unnecessary suffering and ensure the protection of that principle with respect to conflict. More specifically, the moral standards that underpinned the agreement have been interpreted to directly connect to the taboo sentiments opposing both biological and chemical warfare. A US Central Intelligence Agency (CIA 1969a: 3) report stated that the Declaration 'is considered the earliest evidence supporting the contention advanced from some quarters that CBW [chemical and biological warfare] weapons are proscribed by general practice and customary international law'. The belief that certain forms of harm were morally unacceptable expressed within the Declaration, therefore, contributed to the taboos around the poisonous means of warfare even if the connection was not overt within the text itself.

The proscription against biological and chemical warfare associated with the St Petersburg agreement *was* made explicit within the 1874 Brussels Declaration. The Brussels document was negotiated in the wake of the Franco-Prussian War as an initiative of Czar Alexander II of Russia (H. Harris 1973: 11; Mancini 2011: 328). The Czar was reportedly moved 'by humanitarian feeling ... to attenuate the effects and cruelties of warfare' (Bujard 1974: 528). Admittedly, the *New York Times* was less optimistic of the Czar's motivation from a moral perspective. A newspaper article published at the time dismissed the agreement as merely an exercise in Russian self-interest and as designed to promote an expansion in the rights of belligerents at the time ('The Brussels Conference', 1874: 4). Yet the agreement did still endorse a specific restriction on poisonous warfare as a moral consideration. Article 13 of the Declaration ruled against the 'employment of poison or poisoned weapons' and that this mode of aggression was deemed to be 'especially "forbidden"'. Critically, the parties to the Brussels agreement singled out biowarfare as a focus of prohibition during the document's negotiation. Bernard Brungs (1964: 47–8) demonstrates that the threat of bioweapons was raised on a number of occasions in the preliminary debates as a concern within its own right. A working draft of the Declaration had even specified that the agreement's text should ban 'the spreading, in any means whatsoever, of disease on enemy territory' (Brungs 1964: 47). Delegates later suggested, however, that the wording could be simplified with a more general reference to poison and this revised phrasing was adopted within the final version. Consequently, the Brussels document establishes that biological warfare was clearly included within the concept of 'poison' that was employed at this time.

The Hague Conferences

The Brussels Declaration ultimately 'came to nothing' due to a lack of international support (Atlay 1905: 12). Yet the agreement still laid the groundwork for the 1899 Hague Peace Conference ('Peace Conference: England', 1899: 7). The diplomatic negotiations at the Hague produced three conventions, including the 1899 *Convention II with Respect to the Laws and Customs of*

War on Land. Signatories to this Convention were required not to 'employ poison or poisoned arms' as part of Article 23. A further declaration attached to the Convention was also agreed and called on parties to 'abstain from the use of projectiles the sole object of which is the diffusion of asphyxiating or deleterious gases'. In addition to these documents, a similar provision had been proposed to append a clause to the Declaration of St Petersburg that would prohibit 'the use of asphyxiating bombs'. The proposition failed, however, because many of the delegates accidentally left the room when the vote on the policy took place ('Point Won', 1899: 3). The participants were not told clearly that the matter was being decided and they missed the vote by leaving the discussions prematurely. Despite this administrative hiccup, however, the Convention and its associated declaration still comprised a clear mark against poison as a weapon. Significantly, the decision to prohibit poisonous warfare was interpreted at the time as an explicitly moral consideration designed to 'avert unnecessary suffering in war' ('Hague Declarations', 1905: 141). The Convention, therefore, also expressed a recognition that poison weapons were deemed taboo.

The concept of poison employed at the Hague again extended to bioviolence (DeLupis 1987: 212). Eric Merriam (2014: 4–5) claims that the additional declaration's reference to asphyxiating weaponry was explicitly a chemical warfare concern and that the threat of biological arms was only tacitly included within the document. Yet Merriam's interpretation would overlook that the provision still potentially included the delivery of certain biological agents by gas, as well as the wider reference to poison in Article 23 of the Convention. The Article explicitly adopted the wording of the Brussels Declaration, which itself acknowledged the biowarfare risk—as discussed above. Interestingly, disease was something of a feature of the conference. The city's canals resulted in an outbreak of malaria and many of the conference delegates contracted the sickness. The wife of one of the American delegates was even reported to have been removed to Dusseldorf for 'a change of air' as a consequence of the infection ('The Peace Conference', 1899: 7).

The US approved Article 23 but refused to support the additional declaration on asphyxiating shells. An American delegate, Captain Alfred Mahan, had led a charge against the Hague's proscription of these specific armaments at the conference. Mahan argued that he could not see a clear distinction between asphyxiating shells and other forms of military aggression ('Capt. Mahan's Argument', 1899: 7):

> It was illogical, and not demonstrably humane, to be tender about asphyxiating men with gas, when all were prepared to admit that it was allowable to blow the

bottom out of an ironclad at midnight, throwing four or five hundred into the sea, to be choked by water, with scarcely the remotest chance of escape.

(in 'The Peace Conference: Private', 1899: 5)

Mahan further attempted to reason that asphyxiating shells were in fact *more* moral than types of conventional warfare. Mahan (in J. Scott 2000a: 366) stated: 'Such [gas] projectiles might even be considered as more humane than those which kill or cripple in a much more cruel manner, by tearing the body with pieces of metal.' Mahan (in J. Scott 2000a: 375) additionally opined that asphyxiating shells were not 'cruel without being decisive'.[8] While his comment acknowledged that the armaments were horrific in effect, therefore, he maintained that their military effectiveness could outweigh the cruelty in question. Indeed, Mahan (in J. Scott 2000a: 296) elsewhere upheld the pragmatic need to retain all strategic and tactical options in warfare. The conference minutes described his position:

While he is representative of a country which cherishes a keen desire to render warfare more humane, he also represents a nation which may be compelled to wage war, and it is therefore a question of not depriving oneself, by means of resolution hastily reached, of means which later on might by usefully employed.

(Mahan in J. Scott 2000a: 296)

Mahan's opinion, however, was not the norm in respect to the conference. According to the *Manchester Guardian*, his position 'excited a holy horror' in that the other delegates at the Hague still believed that asphyxiating arms comprised an illegitimate means of violence which needed to be controlled ('Parting Reflections', 1899: 8). The rejection of Mahan's position was even expressed by other members of the US delegation itself and his move against the ban was not representative of the wider American view (White 1905: 748–9). US historian William Isaac Hull (1910: 5) would soon after describe Mahan's stance as 'most regrettable'. Admittedly, Mahan did have the support of the United Kingdom (UK) on the matter. UK backing, however, was a case of political convenience. The British agreed to oppose the inclusion of asphyxiating shells at the conference only if the US would object in return to a prohibition on dumdum bullets. An anonymous delegate commented on this deal and the US–UK relationship more generally: 'Blood is thicker than

[8] Mahan was initially quoted in the meeting's minutes as having stated that asphyxiating shells were not clearly 'uselessly cruel' in their effects (in J. Scott 2000a: 296). Mahan requested that the record be changed to not 'cruel without being decisive'.

water, there is no mistake about that' ('The Peace Conference: Asphyxiating Shells', 1899: 7).[9] The UK was willing to put established political alliances first. Yet the British delegation's pragmatism does not automatically mean that they opposed the principle of a restriction on poison armaments. The British refusal to support the provision on asphyxiating shells was a means of ensuring diplomatic gains elsewhere—albeit one which was apparently considered more imperative than upholding the proscription on certain forms of poisonous warfare. Ultimately, the US and UK largely stood alone on this matter (Keefer 2006: 13). Article 23 and the declaration were upheld more widely. Herbert Manisty (1923: 19) noted that, even despite Mahan's position and the US's refusal to sign the declaration, the international commitment to the document's principles had 'never been denounced or repudiated'.

To the extent that Mahan's arguments questioned the bioweapons taboo, therefore, his belief was viewed as an aberration and his position dismissed as highly conditional. Mahan was a long-standing maritime official. He was subsequently accused of misinterpreting the issue of asphyxiating shells exclusively from the perspective of naval war (J. Scott 2000a: 283). Delegates contended that his naval bias both overlooked and misunderstood the threats associated with asphyxiating warfare on land. Yet even Mahan acknowledged that other forms of poisonous warfare were reprehensible—even if he then attempted to use this claim to justify his position in relation to the controversial shells. Mahan (in J. Scott 2000a: 366) asserted: '[T]he use of projectiles of the kind in question can not be considered as being a means which is prohibited on the same ground as the poisoning of waters.' Mahan effectively sought to demonstrate that the shells in question did not fall within the taboo that had developed around alternative means of poisoning within conflict. Shells, then, should not be prohibited on the basis of his distinction. Mahan's comment, however, still recognized that a taboo existed, at least on a conditional basis. Furthermore, other delegates *did* accept that the normative opposition to poisoning water was applicable to asphyxiating shells. Russian delegate, Captain Scheine (in J. Scott 2000a: 366)—who had initially proposed the declaration for negotiation—maintained that asphyxiating projectiles *were* on a par with the contamination of water and that the shells were similarly as 'barbarous in character' as the latter.

The Hague Conference's final agreement was presented as an expression of moral concern. A letter from the Pope was read out at the end of the debates in

[9] Hull (1909: 84) also commented on the US–UK agreement: 'Thus, although "dum-dum" bullets were originally British chestnuts, and although they have been pronounced worm-eaten by the British themselves, our United States government, largely thanks to Captain Mahan's reactionary principles, still insists on pulling them out the fire.'

support of the agreement's principles ('Close of the Hague Conference', 1899: 5). The letter demonstrates both the Convention's commitment to moral standards as a core objective and the religious legitimation of those principles as such. The taboo was reasserted at the subsequent 1907 Hague Conference. Delegates argued here that there was no need for any revision, or even a revisiting, of the provisions against biological and chemical weapons established in 1899. It was maintained that any return to the issue would be 'out of order' and disrespectful to the integrity of the international norm against poison warfare that had been expressed at the first conference (J. Scott 2000b: 105). Admittedly, the moral aims of the conferences were treated with some contemporary scepticism in certain quarters. A *New York Times* commentary said at the commencement of the 1907 proceedings:

> When, as one cynical philosopher has observed, all of the women are dead and all the wine drunk up, then may the world look for universal peace, then may the lion lie down with the lamb, and, as still another cynic has added, with the lamb inside.
>
> **('On the Eve', 1907: SM8)**

While there was then a taunt at the time that 'being summoned as Peace Conferences they turned themselves into War Conferences', however, both Hague events were still ultimately seen to be instrumental in establishing and codifying key ethical norms pertaining to war (Lawrence 1908: 83). These norms included the direct proscription of biowarfare as a moral concern.

World War I

The Hague provisions were apparently insufficient to prevent the use of bioweapons during World War I.[10] A Stockholm International Peace Research Institute (SIPRI 1971b: 216) investigation reports that German forces infected Allied horses and cattle with glanders and anthrax during the conflict. Germany was also alleged to have spread plague in St Petersburg in 1915, contaminated wells with corpses at the Somme in 1917, and distributed fruit, chocolate, and children's toys infected with lethal bacteria throughout Romania (Smart 1997: 16). Extensive question marks still hang over the validity of these allegations; although the evidence concerning the German assaults on animals has been considered compelling. Yet it is notable that this specific form of attack would represent a strike against

[10] George Bunn (1969: 376) argues that the Hague's 'limited' language meant that the agreement had little impact on chemical and biowarfare during World War I.

non-human targets. The choice of target potentially qualifies the violation to the extent that the attacks may not have been viewed as entirely incompatible with the taboo on biowarfare in that they did not seek to directly harm humans.

Indeed, Germany had elsewhere expressed a commitment to the taboo. The 1902 *War Book of the German General Staff* (2005: 18) explicitly prohibited the use of poison within warfare, including a direct rejection of 'the propagation of infectious diseases'. Mark Wheelis (1999b: 38) and Toby Ewin (2020: 284) separately recount that a suggestion was put forward in 1916 for Germany to use Zeppelins to drop plague on enemy ports with the aim of infecting rats and so cause an epidemic. Yet both analysts also highlight that the plot was rejected. Specifically, the plan was not adopted due to ethical concerns that reflect the taboo. The German Surgeon General's office replied to the plan's proposer stating the following: '[A]ll respects to your courage and patriotism, but if we undertake this step we will no longer be worthy to exist as a nation.' Taboo expectations of civilized conduct trumped military advantage in this case. Ewin (2020: 284) additionally reports that there was a further proposal from the German embassy in Madrid the same year to use *cholera bacilli* to contaminate the rivers near the Portuguese border with Spain. The strike would have been enacted in response to Portugal entering the war. Yet again, however, the suggestion was not approved.

Chemical weapons, however, were clearly employed during World War I. The horror of chemical warfare inspired a wider shift in both political and public perceptions of the threat which cemented the taboos against chemical and biowarfare. As a partial caveat to this statement it is acknowledged here that the response to chemical warfare was mixed. Revulsion was not universal. Analysts such as J.B.S. Haldane (1925) even argued that chemical weapons were *more* humane than conventional forms of aggression. Yet the experiences of World War I still exacerbated and entrenched a widely adopted viewpoint that chemical warfare was intolerable. The stance against chemical warfare was expressed within taboo terms in that the threat was seen to comprise an immoral act which demanded prohibition. A *British Medical Journal* commentary stated of Germany's use of chemical agent:

> The use of poison gases, which had been rejected several times by the British Government on the score of inhumanity, was expressly forbidden by the Hague Convention of 1899, to which Germany was a party. The deliberate violation of this agreement by the Germans in 1915 is a particularly flagrant addition to the lengthening tale of their acts of barbarism.
>
> **('The Use of Asphyxiating Gas', 1918: 169)**

Chemical violence was further deemed to constitute a normative transgression that threatened the very stability of society itself due to its immoral nature (Jones 2015: 129). Yet taboo sentiment was not limited to chemical warfare only. Normative opposition also extended to biowarfare, whereby these weapons were also classified as a poisonous act of conflict. The association of chemical and bioviolence was exacerbated at this time as a consequence of another key international event: the Spanish Flu (Steinbruner 1997: 85). The pandemic reignited fears as to what could occur if disease was harnessed as a weapon. The conflation of the use of gas in the trenches with the infectious global outbreak under the umbrella of poison warfare expanded contemporary concerns around chemical aggression to biowarfare and so further highlighted the threat of disease as an illegitimate form of aggression.

The concept of poison was again employed within Article 171 of the 1919 Treaty of Versailles. The agreement banned 'asphyxiating, poisonous, or other gases' in respect of Germany as part of a wider set of measures to preclude extensive German rearmament after World War I. Despite the core focus on containing German power, however, the prohibitions were also seen to directly confirm the pre-existing norms against poison warfare (Buell 1922: 207; MacLeod 2017: 196). The same reference to poison also appeared within the 1922 Washington Treaty. America had not been a signatory to many of the previous agreements against biological and chemical warfare outlined above. The US, however, would comprise a leading participant in this *Treaty Relating to the Use of Submarines and Noxious Gases in Warfare*. The conference was initially set up to establish export prohibitions on offensive gases only. It was argued, however, that a sole concentration on gas would not prevent chemical and biowarfare more widely or the manufacture of such agents for war purposes. Consequently, the conference subsequently sought a more extensive ban. Delegates concurred that a more limited prohibitive regulation would be highly problematic, if not impossible, in terms of an agreement's workability and would not fully address the moral imperative to halt further modes of poisonous violence ('Parley', 1925: 14). The treaty deliberately drew upon the language established at Versailles where this wording had already attained international approval (Moore 1972: 432).

The Geneva Protocol

The Washington Treaty would not come into force as France refused to ratify the agreement over the restrictions on submarines. The agreement was, however, a precursor to an even more critical aspect of the biowarfare taboo's codification: the 1925 Geneva Protocol. The US again adopted a leading role in the proceedings. Secretary of State, Frank Kellogg (1929: 73), publicly stated that it was America who had taken 'the initiative' in bringing the Protocol discussions to the international table. Moreover, there was a serious concern at one point in the negotiations that the Geneva conference would not produce an agreement. US President Calvin Coolidge stepped up and authorized the head US delegate, Congressperson Theodore Burton (R: Ohio), to offer to host a further conference if needed in order to ensure that the Protocol happened (Burton 1925: 56). Consequently, the US again led the way. In announcing the offer of further negotiations, Burton quoted Coolidge on the moral need for such an agreement in an expression reflective of the taboo: 'One of the greatest attributes of all people is the desire in spite of all things to do the right thing' (in 'Arms Conference', 1925: 6). Burton personally supported a ban on all poisoning armaments. A US Operations Research Office (ORO 1959: 198) report reveals that he had persuaded Coolidge to support a treaty preventing international trade concerning these weapons in what would effectively comprise a non-proliferation agreement.

A second conference was not required. The Protocol was approved and established an international prohibition on both biological and chemical warfare. The agreement was highly conditional, however, and did not comprise an outright ban. The Protocol's terms were limited in that the treaty forbade chemical and bioviolence only within the context of war (where what precisely constituted a state of 'war' was not clearly articulated) and parties were permitted to continue with even offensive research and weapons stockpiling. Many signatory states also 'added' to the Protocol through a number of controversial reservations. One reservation limited the application of the Protocol only to actions concerning other parties to the agreement. No restraint was required to be shown towards those who had not officially approved the document. Moreover, a further reservation permitted observing states to respond in kind to the use of the proscribed weapons against them. Consequently, the Protocol essentially constituted a no-first-use arrangement as opposed to a full ban or non-proliferation effort. In addition, there was a problematic lack of clarity as to whether the agreement applied to any weapon developed *after* the Protocol's negotiation. It was argued that the document's conclusion in 1925 could be interpreted as a 'cut off' to future advances in weapons technology, which would not then be included within the terms of the prohibition (Neilands 1970: 210; DeForest 1990: 24). Would the codified controls apply to new weapons that were developed subsequent to the agreement, where this situation could be seen to represent an objection to all means of poisonous warfare generally as a normative rule? Or was the Protocol specific only to those armaments available at the time? The document was vague on this point.

Brungs (1964: 66) has dismissed the inclusion of bioweapons at Geneva as merely an 'afterthought'. Similarly, Ann Synge (1969: 260) claimed that biological armaments were simply 'thrown in for good measure because nobody had any'. Yet the archives show that bioweapons were incorporated within the Protocol as part of a dedicated effort to do so. Head of the Polish delegation at Geneva, General Kazimierz Sosnowski, had submitted a proposal outlining an explicit reference to prohibitions on biowarfare at the committee stage of the negotiations. Poland had expressed a vested interest in the Protocol. The Polish delegates viewed the agreement as a protection from the potential threat posed by the Great Powers ('General', 1970: 105)—a concern that would be more than realized in 1939. In light of such a threat the Polish delegation believed that the Protocol needed to specifically address the issue of biological aggression ('General', 1970: 111). A general ethos of restriction on poison warfare was considered valuable but insufficient without a clear and separate constraint in respect of biowarfare. Only then could the advantages of unconventional warfare which may be exploited by the Great Powers be neutralized.

Yet the Polish demand was not simply a practical calculation of security. Delegates constructed their case against bioweapons in taboo terms. In one example, the Polish amendment was approved but accidentally missed off the text put forward for the final discussion. Poland's delegation called emphatically for the wording to be reinstated and specifically asserted that the change should be enacted on the basis that bioweapons could create even greater horror than chemical weapons ('Agree on Search', 1925: 6). As such, it was the taboo terror associated with biological military devices that was cited as the core reason for these armaments' inclusion. Sosnowski (in 'General', 1970: 112) himself elaborated further on this taboo idea, saying: 'The actual results and also the *moral* consequences of such a [biological] war would therefore be terrible' [emphasis added]. His justification for reinstating the explicit prohibition against biowarfare was again, therefore, an intrinsically taboo argument. The US delegation publicly accepted the Polish amendment and repeated these taboo sentiments in doing so. American officials expressed their agreement that biowarfare should be stigmatized as 'so revolting and so foul that it must meet with the condemnation of all civilized nations' (in Brungs 1964: 65–6). The final version of the Protocol similarly declared that 'the use in war of asphyxiating, poisonous or other gases, and of all analogous liquids, materials or devices, has been justly condemned by the general opinion of the civilised world' (*Protocol for the Prohibition of the Use in War of Asphyxiating, Poisonous or Other Gases, and of Bacteriological Methods of Warfare*), 1925). The Protocol further appealed to the 'conscience and practice of nations', where the entreaty to conscience again connects to the moral sentiment of the taboo.

Consequently, the Protocol reinforced a taboo against poison warfare that had been maturing in terms of its codified status for many centuries. Yet the document also represents a crucial step in the bioweapons taboo's development in that the Protocol explicitly highlighted the threat of biowarfare. Parties commit to a specific 'prohibition to the use of bacteriological methods of warfare'. Prior agreements had subsumed the biowarfare threat within a more general reference to poison. Bioviolence was now distinguished as a concern within its own right—albeit as part of an agreement that addressed both biological and chemical arms control simultaneously. The clarification is additionally relevant in that the previous conflation of bioweapons with their chemical counterparts had been seen to be problematic. Major William Neinast (1964: 22) referred to the 'inexcusable confusion of biological warfare with chemical warfare' before the Protocol was enacted. The agreement, therefore, formalized a key and necessary distinction between two different forms of violence. Furthermore, the bioweapons taboo—as a real, measurable, and specific international ruling against biowarfare—was now unquestionably evident.

Customary prohibition

US ratification of the Geneva Protocol was considered to be in the bag. America had led at both Washington and Geneva. US officials had repeatedly declared that the country 'condemned the use of inhuman and barbarous methods [of war] … since the Hague convention' and the Protocol was 'in accordance with our [American] settled policy' (Ulm 1926: 5; Moore 1972: 433). The agreement had gone forward to the Senate for approval following an 8–1 vote in favour at the Senate Foreign Relations Committee (SFRC) stage. Coolidge and Kellogg were, therefore, 'overconfident' that ratification was a mere formality (Zablocki in USCong 1969a: 76). Eyes off the ball, anti-Protocol commercial and bureaucratic interests intervened to derail the process. The situation was exacerbated by the extensive length of time allocated to the internal ratification discussions: 18 months from the Protocol's signing to the ratifying vote. The delay allowed various interests against the Protocol the time to mobilize into a unified and effective pressure group (ORO 1959: 199; Jones 1980). The group attracted strong support within the US government and military structures. Senators, members of the armed services and War Department, as well as the new Chemical Warfare Service (CWS)—the US domestic military organization responsible for chemical and biowarfare, founded in 1918—were willing to back the movement opposing the Protocol. These actors believed that they had been unfairly excluded from the ratification process and now 'felt completely at liberty to attack the treaty' (Bunn in USCong 1969a: 49–50).

Other interested organizations outside of government and the military also joined the fray. These groups included the American Chemical Society (ACS), Association of Military Surgeons, as well as veteran associations such as the American Legion and the Veterans of Foreign Wars ('Chemists Denounce', 1925: 12). ACS President, James Norris (in 'Pershing', 1926: 3), 'charged that the opponents of poison gas warfare were insincere, that their advocacy of the Geneva Protocol was based on misinformation, and that the widespread feeling against gas warfare was a relic of war hysteria and propaganda'. Editorials in publications such as *The Journal of Industrial and Engineering Chemistry* similarly accused the US government of prioritizing public opinion over considerations of national and military security ('Popularity versus Security', 1922: 178). Such articles dismissed taboo concerns over poisonous warfare as mere 'popular sentimental feeling' ('The War on Chemical Warfare', 1922: 98).

The campaign against the Protocol was a success. SFRC Chair, Senator William Borah (R: Idaho; nicknamed 'the Lion of Idaho'), withdrew the treaty from Senate consideration once it became clear that the agreement simply did not have sufficient votes in favour. One of the Protocol's key supporters, General John Pershing—who had personally experienced chemical warfare while fighting in World War I—wrote to Borah expressing his disappointment at the agreement's withdrawal. Pershing (in 'Our Position', 1927: 5) said: 'I cannot think it possible that our country should fail to ratify the protocol ... chemical warfare should be abolished among the nations as abhorrent to civilization.' The Protocol technically remained with the SFRC until 1947. The agreement was then officially dismissed in a bureaucratic housekeeping move by a new Chair of committee, Senator Arthur Vandenburg (R: Michigan). Significantly, however, the anti-Protocol campaign was specifically an attack against the prohibition of *chemical* weapons. Opposition to the agreement did not comprise a rejection or contestation of the bioweapons taboo per se. At no point was the objection explicitly aimed at questioning the normative expectations against bioviolence. The Protocol's failure was also not a consequence of a wider lack of popular or political opinion against biological warfare, which was well established at this time. Indeed, the strong public support for the norm against poison warfare comprised a key reason why actors believed that there would be no complications relating to ratification and why there was an absence of attention paid to ensuring that the motion would pass.

Furthermore, the withdrawal of US support did not inhibit the Protocol's customary impact. The agreement not only contributed to the codification of the bioweapons taboo; the document additionally established the taboo as an international custom, that is, the Protocol was considered applicable even to those actors who had not officially ratified the treaty (e.g. Moore 1972: 449; DeForest 1990: 24). Consequently, the US could be deemed fundamentally bound by the Protocol's principles despite its non-party status (Geissler, van Courtland Moon, and Pearson 1999: 257). Admittedly, it has been argued that the removal of American support for the Protocol undermined its claim to customary standing (Stein 1972: 281). US officials, however, would consistently state that they recognized and adhered to the Protocol, and to the taboo which underpinned the agreement. The US's commitment to the Protocol will be discussed more in later chapters. For now, this chapter refers to a quote by American Professor of Microbiology, Sydney Rubbo. Rubbo (1968) said seven years prior to US ratification of the Protocol that the agreement was:

> [R]egarded as internationally binding and might also be regarded as the universal Hippocratic oath—'Never will I administer poison when asked to do so, nor will I suggest such a course'. ... [T]he Protocol should stand as the 11th Commandment of human behaviour.
>
> **(Rubbo 1968: 23–4)**

The Protocol's customary normative impact was so extensive that this influence has been considered even more instrumental in respect of American political action than any calculation of military utility or strategy. Julian Perry Robinson (1981: 520) expressed that the Protocol specifically reinforced types of institutional and normative restraint *over* military considerations. He further commented: 'It is at this level, arguably, that the Geneva Protocol exerts its greatest influence.'

The Protocol's customary status has been contested on the grounds of its conditionality and the reservations attached. It is claimed that the various limitations established in respect of the agreement means that the Protocol failed to establish a customary provision against biowarfare in that the document was too qualified and partial to do so (O'Brien 1962: 30). Neinast (1964: 30) makes the even stronger argument that the reservations not only failed to confirm the presence of customary status but actively undermined it: 'Thus if biological warfare were prohibited by custom, the Geneva Gas Protocol did not codify custom; it *repealed* the custom' [emphasis added]. The taboo is, in effect, an outright ban on bioviolence. The Protocol then placed legal caveats on that ban, for example that weapons use was permitted in certain circumstances. Consequently, says Neinast, the Protocol can be interpreted as having made the normative rejection of biowarfare into a lesser proscription. Other analysts take this line of argument even further still. McCauley and Payne (2010: 12) assert that the reservations—such as that which permits a retaliation in kind—have in fact encouraged bioweapon development and use. Actors were inspired to construct bioweapons as a means of ensuring access to a retaliatory capacity. The authors then state: 'Many nations used this opening [like-for-like response] as an entryway to develop bioweapons.'

Yet it is still the case that the Protocol comprised a crucial step in establishing the bioweapons taboo. The agreement officially reaffirmed and gave a new global voice to the taboo repulsion felt towards disease and its weaponization. The US was committed to those principles even if it did not officially ratify the treaty itself. In addition, the Protocol comprised a fundamentally moral expression in support of the taboo. Francis Boyle (1985: 1182) asserted that the agreement should not be reduced to a practical arms control agreement or simply a response to chemical warfare as employed during World War I. Boyle maintains that the Protocol should be considered as an intrinsic contribution to a wider push for a more humanitarian warfare. Furthermore, he argues that the normative statement made by the Protocol in this respect was, and remains, so robust that it 'overcomes' the contentious problems associated with the Protocol's conditionality and reservations. The normative value of the agreement in promoting the taboo negates its limitations. A similar argument is proposed by Westing (1985b: 117), who asserts that the Protocol's 'strength' does not lie within its specific provisions but rather 'from the widely shared antipathy and even revulsion toward that particular means of killing one's fellow humans; and, as a corollary, from the shame attached to using such weapons.'

It is within this context of the Protocol's normative expression of the bioweapons taboo that the agreement would become viewed (prior to the BWC) as the 'cornerstone' of biowarfare prohibition and an essential factor in future developments relating to biological violence beyond 1925 (SIPRI 1971a: 18). The Protocol was a critical and necessary aspect of the taboo. For, while the taboo was not yet robust with respect to the Protocol itself, the progression towards the taboo's greater normative strength would explicitly build on and seek to fully realize the customary intent that had been recognized, expressed, and decreed at Geneva. This book now analyses this progression across the coming chapters.

4

Construction

The US Bioweapon Programme

When the US instigated its bioweapon programme in 1942 it was uncertain whether biological aggression constituted a realistic or effective option. Strategic understanding of biowarfare had been largely dependent on a pessimistic study by Major Leon Fox of the US Army Medical Corps published in 1933. Fox held that bioweapons were highly problematic: it was unclear whether biological agents could withstand weaponization; delivery was extremely difficult; there was the issue of the 'boomerang' effect (in that any disease released into the wild could ultimately rebound upon the attacker); and bioweapons could be mitigated against by measures such as mass immunization and protective clothing that could vastly reduce their destructive impact. Fox (1933: 189) concluded that biowarfare was a fantasy: 'Bacterial warfare is one of the recent scare-heads that we are being served by the pseudo-scientists who contribute to the flaming pages of the Sunday annexes syndicated over the Nation's press.' Other strategists repeated his doubts. Major General Harry Gore Bishop (1934: 5), the US Army Chief of Field Artillery, rejected the use of biowarfare on the basis that it was uncontrollable. Regarding the concerns around the boomerang effect, Bishop declared that these arms were 'no respecter of persons' as there were no means 'to limit germs to one side of No Man's Land'.

Consequently, it was argued that biowarfare did not pose a major threat. A 1934 article in the *New York Times* openly ridiculed warnings by the Chief Medical Officer of the French Army, General Romieu, concerning the risks of bioweapons. Romieu had claimed that bioviolence comprised an inherently disgusting threat which lay in moral opposition to the otherwise chivalrous nature of the military.[1] He was accused of seeing 'bogies' and his concerns dismissed as ill-informed conjecture, egged on by the more tabloid aspects of the media and 'army officers who are better qualified to express an opinion on tactics than on bacteriology' ('Germs in the War', 1934: XX6). US Lieutenant

[1] The article noted that Romieu had specified that the French military in particular was especially chivalrous, 'as he may be pardoned for stressing' ('Germs in the War', 1934: XX6).

The Biological Weapons Taboo. Michelle Bentley, Oxford University Press. © Michelle Bentley (2023).
DOI: 10.1093/oso/9780198892151.003.0005

Colonel James Simmons received a similar response in 1937. Simmons was a member of the Medical Corps and issued a military report cautioning that an enemy could deliberately employ insects to spread diseases such as yellow fever against the US. Simmons' assessment, however, attracted only minimal interest (Lockwood 2009: 142). Set against a wider background of American isolationism—whereby the military was regarded as 'a pesky necessity during peacetime'—and a reticence towards financial expenditure following the Great Depression, arguments that biowarfare constituted a credible threat failed to gain traction within both political and military circles (S. Harris 2002: 202). Bioweapons were not even on the agenda let alone a priority concern.

Perceptions of biowarfare at this time could be interpreted as entirely predicated upon pragmatic calculations of viability. If biological agents could not be weaponized to a point at which they could make a feasible contribution to military strategy, then non-use and any failure to incorporate bioweapons within strategy was a consequence of practicality. More specifically, avoidance of bioweapons was not a consequence of normative restraint. Bishop (1934: 19) explicitly rejected the idea that the resistance to biowarfare should comprise a moral decision. Morals were a luxury that the prudent strategist could ill afford, especially under conditions of active warfare. Bishop acknowledged, however, that those morals existed. The Major General accepted that bioweapons were deemed 'tabooed' by those in 'civilized nations' (Bishop (1934: 19). While he disapproved of letting such moral considerations interfere with the business of the battlefield, he still recognized the taboo's existence. Critically, Bishop also believed that this taboo had the capacity to significantly influence thinking on military issues. In line with the customary status of the taboo attached to the Geneva Protocol (discussed in Chapter 3), it was now accepted that biowarfare was not merely an issue of practicability but also of stigmatization and moral feeling.

Prelude to a bioweapon programme

The clash between Krasner's 'gun to the head' logic and normative dynamics characterized the US bioweapon programme's development. In terms of the desire for security, World War II can explain why America came to initiate an extensive and costly programme less than a decade after Fox had ruled bioweapons out. The war exerted pressure to secure the US by any means necessary. The programme was also a response to evidence that other states were constructing bioweapons (Geissler, van Courtland Moon, and Pearson

1999: 259; Roffey, Tegnell, and Elgh 2002: 452; Tucker 2002: 109; McCauley and Payne 2010: 12). Japan had launched a major programme under the leadership of Surgeon General Shiro Ishii at the now infamous Unit 731. The unit was a secret facility dedicated to weapons testing and truly barbaric human experimentation.[2] Ishii believed that the Geneva Protocol was a sign that biowarfare was viable and valuable (Regis 1999: 11). The move to conditionally outlaw biological arms demonstrated that these must be weapons of the highest calibre. If not, why bother to proscribe a useless device? The Japanese programme came to the formal attention of US officials in 1939 when one of Ishii's associates, Dr Naito Ryoichi, was reported to the Federal Bureau of Investigation (FBI). Ryoichi had repeatedly attempted to obtain yellow fever (the same disease that Simmons had identified as a potential threat only two years before) from the Rockefeller Institute in New York, including offering an employee $3,000 as a bribe (G2 1941).

Germany was also a threat. The state had been accused of developing a biowarfare programme prior to World War II. In the same year that Fox published his seminal analysis, Germany supposedly carried out biological tests on the Paris Metro and London Underground. These claims were the subject of a sensational news report by Henry Wickham-Steed (1934). Yet the allegations were met with scepticism at the time. *The Observer* newspaper maintained it was 'unlikely that the experiments described in Mr. Steed's article ever really took place' ('Germs as War Weapon', 1934: 18).[3] Similarly, there was little concern in 1937 when *Deutsche Wehr* published a German military document that appeared to incorporate biowarfare as an official strategy ('Germans Studying', 1937: 14). While the document actively promoted biowarfare, analysts again cited the same ideas espoused by Fox to claim that such tactics remained unfeasible. One US newspaper drew on the concern of the boomerang effect to argue:

> If microbes are ever used in combat they will have to be as finely disciplined as the German Army. Instead of flying hither and thither in their present vagabond way, they must goosestep straight for the enemy, shoulder to shoulder, as it were, and lodge in his lungs, his food, his water supply, his milk.
>
> **('Bacterial Warfare', 1937: 22)**

[2] This study will not analyse the Japanese biological programme in depth, but for further details see Sheldon Harris (2002) *Factories of Death: Japanese Biological Warfare, 1932–1945, and the American Cover-up* and Jeanne Guillemin (2017) *Hidden Atrocities: Japanese Germ Warfare and American Obstruction of Justice at the Tokyo Trial.*

[3] Modern assessments of the allegations repeat this scepticism; see Martin Hugh-Jones (1992).

The article concluded: 'It is not likely that army officers will take the Deutsche Wehr's inquiry any more seriously.'

World War II challenged this perception. Politicians in the US and UK expressed their anxiety that Germany would use bioweapons as a means of last resort as the war progressed ('Nazis Held Ready', 1941: 14). New intelligence ostensibly confirmed this concern. An American military attaché in Switzerland reported in May 1941 that Germany was using French facilities (specifically, the Koch Foundation Laboratory outside of Paris) to weaponize botulinum toxin (Buxton 1943). The claims played into wider fears that the Nazis had expanded their rocket capability, which would increase the chance and viability of biowarfare (CIA 1944; Lovell 1944). The quality of this evidence was extremely poor. Within the context of World War II, however, many US strategists were inclined to take the threat seriously. Seymour Hersh (1968a: 12) recalls a 1957 speech by Major General Brock Chisholm, Director-General of the World Health Organization from 1948 to 1953. Chisholm recounted the Allies' dread that German rockets would be biological: 'When the V-1 bomb went off with a big bang, showing that it only contained normal high explosives [not biological agent], the general staffs all heaved an immense sigh of relief.'

In response to this perception of biological threat there was an 'underlying assumption' that the US should develop all retaliatory capabilities available to them (van Courtland Moon 1999a: 252). Moreover, it was simply inconceivable that an Axis power should gain a head start on the US in relation to any aspect of military security. Consequently, Rexmond Cochrane (1947: 31) stated that the US 'had no alternative but to prepare against a possible [biological] attack'. The US's own allies were also developing a biowarfare capacity; for example, the UK programme at Porton Down. Allied acquisition of bioweapons did not pose the same threat as an Axis capability. Yet US actors still felt that they could not be left behind in the biowarfare race—especially as it now appeared that they would be the last to the starting line in doing so.

Taking offence

The first step in acquiring bioweapons was an August 1939 CWS report entitled *Technical Study No. 10: Bacteriological Warfare Possibilities* (CWS 1939). The analysis outlined a list of biological agents that could potentially be weaponized, for example smallpox, yellow fever, and bubonic plague. The CWS essentially established that there was at least a discussion to be had concerning a US bioweapons capability by identifying these agents—in that the report had now hypothetically set out what such a programme might look like. This first step, however, was highly speculative. The report was by no means an enthusiastic endorsement of biowarfare as a reliable or usable strategy, contra Fox. While the assessment opened up a dialogue on a possible programme, this was only the vague suggestion that biowarfare might not be ruled out entirely and more detailed study on the matter was necessary. A US Department of the Army (DoA 1977: 1-1) analysis described the situation as a case where '[s]ufficient doubt [of bioweapons' effectiveness] existed so that reasonable prudence required that a serious evaluation be made as to the dangers of a possible attack'.

In putting biowarfare on the table, however, *Technical Study No. 10* was adamant on one key point: any future programme should not be offensive. Bioweapons would be limited to defence only. Simmons (1941) voiced the same sentiment in a memorandum for the Special Assistant to the Secretary of War, Harvey Bundy, reiterating his 1937 position. Simmons asserted that offensive bioweapons research was acceptable only to the extent that it contributed to a defensive programme. That is, actors needed to understand the offensive in order to create a defence against this. Bundy responded to the memo by authorizing a consultation in August 1941 comprising officials from the Office of the Surgeon General, CWS, National Research Council, G-2, and the Office of Scientific Research and Development (Brophy, Miles, and Cochrane 1959: 103). The group recommended a further investigation into the viability of biowarfare, which the Secretary of War, Henry Stimson, entrusted to the National Academy of Sciences (NAS). A secret civilian committee of 12 scientists known as WBC was formed to carry out the analysis and first met in November 1941.

Nine days after the committee's initial meeting the US received intelligence that Japan had carried out a biowarfare strike using plague against the Chinese village of Changteh ('Invisible weapon', 1942: 22). Ten days after that, Pearl Harbor was attacked. Both events pushed the committee towards a more pro-biowarfare stance. The WBC concluded in its first report issued in February 1942 that bioweapons use was 'distinctly feasible' (NAS 1942a). The committee effectively rejected the arguments that bioweapons lacked military utility which had previously characterized the debate. The members accepted that there was still a question mark hanging over the exact viability of biowarfare. They argued, however, that the destructive potential associated with bioweapons was sufficient to warrant concern. It was on this basis that they recommended initiating a comprehensive biowarfare programme:

> The value of biological warfare will be a debateable question until it has been clearly proven by experience. Such experience may be forthcoming. The wise

> assumption is that any method which appears to offer advantages to a nation at war will be vigorously employed by that nation. There is but one logical course to pursue, namely, to study the possibilities from every angle, make every preparation for reducing its effectiveness, and thereby reduce the likelihood of use.
>
> **(NAS 1942a)**

A further report in June stated: 'The best defense for the United States is to be fully prepared to start a wholesale offensive whenever it becomes necessary to retaliate' (NAS 1942b). The committee added with emphasis that 'the best defense is offense and the threat of offense'.

Edward Regis (1999: 21) portrays the WBC recommendations as a case whereby the group 'had warmed to their subject and were now firm believers in this new and exotic form of waging war'. Despite the reference to offence, however, the WBC's approach still prioritized the defensive. The WBC only promoted offensive elements within the context of their necessity to defence. However enthusiastic the members were about ensuring a US biowarfare capability if provoked, purely offensive action was never constructed as a strategy within its own right. Consequently, the roots of the US programme were founded on a fundamental belief in non-use except under the direst of circumstances. The very idea of a biowarfare programme was instantly associated with a need to limit military policy to highly constricted modes of weapon use. Actors 'knew' that this was not a weapon that could be developed entirely outside of normative expectations. The fetishized value of biowarfare was apparent: this was a weapon too important, too hallowed, and too disgusting to be used offensively. The taboo, therefore, influenced political and military thinking as the US turned towards a bioweapon programme in the sense that the proposed conception of use was (1) hypothetical and (2) significantly qualified.

First do no harm

The US commitment to a defensive strategy is not the sole evidence of taboo. Simmons (1941) had also recognized an ethical dilemma pertaining to the Hippocratic Oath:

> [S]ince the primary function of the Medical Department is to preserve life rather than to destroy it; should it be deemed advisable to develop facilities of this type of [offensive biological] warfare, this should be accompanied through the activities of some branch of the services other than the Medical.
>
> **(Simmons 1941)**

Morally speaking, medical actors could not make biological bombs. They had vowed not to as a professional commitment. The same issue influenced the WBC's formation. Bundy's consultative group did consider that future investigations might at least consider the possibility of an offensive biowarfare strategy. Yet the members applied a caveat that any issue of offensive use could not be assigned to the remit of a medical body as a point of moral concern. Admittedly, the matter was dealt with in less-than-entirely moral terms. The consultation recommended that two civilian committees be established to discuss US options on bioweapons development: one concerned with the defensive (that would report to the CWS) and one with the offensive (that would report to the Army and Navy). The two committees, however, were to be considered a single entity; a solution that Stephen Endicott and Edward Hagerman (1998: 28) describe as 'hypocritical'. Despite this morally questionable work-around, the taboo—manifest here as opposing the immoral destruction of life through biological means by those ethically committed not to do so—again shaped policymakers' understanding of what a programme would mean and the ideational dynamics that underpinned it.

The civilian nature of the WBC also demonstrates the influence of the taboo. WBC was set up as a civilian body *explicitly* to dispel any association with offensive military elements. The committee's very name is telling. It is uncertain as to what the acronym referred. A US DoA report states that WBC stood for the 'War Bureau of Consultants' (Brophy, Miles, and Cochrane 1959: 103). One of the report's authors, however, had claimed earlier that the WBC reference was a 'misleading inversion of the letters "BW [i.e. bioweapons]"', which he interpreted as a playing down of the connection to a pariah weapon (Cochrane 1947: 34).[4] Similarly, Endicott and Hagerman (1998: 29) argue that the group was initially named the 'Bacteriological Warfare Committee', which was then amended to WBC in a 'deliberate obfuscation' designed to rule out any perception that it was engaged in an offensive remit. It is unclear which version is correct. Yet all these interpretations demonstrate the same reticence towards (1) openly declaring that the committee was concerned with biowarfare policy and (2) associating that policy with offensive means of aggression, per the abhorrence of the taboo.

Normative restraint is further evident in the official decision to start a biowarfare programme. Stimson was not exactly known for introversion in his military beliefs. Yet when he submitted the WBC recommendation in favour of a US biowarfare capability to President Franklin D. Roosevelt for approval, his accompanying memorandum was hardly emphatic in terms of

[4] Harris (2002: 206) similarly claims that 'WBC' comprised a transposition of BWC but says that it is unclear whether this was deliberate.

his support (Stimson 1942). Stimson ostensibly bemoaned that bioweapons were an unhappy necessity and called them a 'dirty business'. His conceptualization of bioweapons as dirty constitutes a model expression of the taboo. The reference reflects both the core ideas of disgust and contamination, as well as Douglas' specific conception of dirt as intrinsic to the notion of a taboo itself.

Critically, Stimson did not embody some perfect conduit of taboo thinking. He had rejected suggestions that the US should consider itself customarily bound by the terms of the Geneva Protocol following Pearl Harbor. Stimson (in Guillemin 2005: 59) felt that tying the US's hands on chemical and biological aggression could 'through introduction of domestic, political and moral issues, impede our preparation, reduce our potential combat effectiveness and be considered, by our enemies, an indication of National weakness'. Yet Stimson still did not promote biowarfare as a positive or an offensive choice. He also supported the preference for a civilian-led programme in order to preclude concerns that bioweapons development would comprise anything other than defensive. His justification for doing so again echoed the taboo: 'Entrusting the matter to a civilian agency would help in preventing the public from being unduly exercised over any ideas that the War Department might be contemplating the use of this weapon offensively' (Stimson 1942).

The US bioweapon programme's central governing authority was named the War Reserve Service (WRS). The Secretary established the body under the auspices of the Federal Security Agency (FSA). The FSA was a New Deal welfare body concerned with issues of public health and social security. Roosevelt (1939: 252) had created the organization with the mission 'to promote social and economic security, educational opportunity and the health of the citizens of this Nation'. As well as the public relations benefits of comprising a civilian (as opposed to a military) institution, therefore, the FSA also had a clear positive association with health. Not least where biowarfare has been described as 'public health in reverse', the FSA offered a legitimizing finesse. An association with the FSA connected the biowarfare programme with perceptions of the healthy—and explicitly that which is not taboo. Yet the construction of the WRS in more palatable terms was not merely a case of public presentation. The WRS' civilian function initially constituted a real restraint on the more aggressive ambitions of the programme. WRS was limited in that it acted only as a coordinating body. Control over actual bioweapons development lay instead within the remit of the CWS. The WRS committee, however, is argued to have overseen policy in ways that stymied CWS plans for more 'elaborate facilities for the development of materials and equipment and for field testing' (Cochrane 1947: 74). Admittedly, the bioweapons programme would still be considerable in terms of its size and function. The CWS would, moreover, eventually force the WRS out in 1944. Under full CWS control, the programme 'was accordingly stepped up' and would include a greater emphasis on offensive research (DoA 1977: 1–2). Yet despite this hard-line influence from the CWS, the WRS still comprised a civilian and non-offensive restraint on what could have been an even more extensive enterprise.

Building the biological bomb

The US programme was set up with an initial budget of $200,000 courtesy of the President's Special Emergency Fund (Bernstein 1987: 117). The programme outgrew its first home at Edgewood Arsenal in Maryland only a year later and construction commenced on a major new complex in the same state, Camp Detrick (later renamed Fort Detrick). More facilities followed as the programme expanded, including field test sites at Horn Island in Mississippi (although testing here was restricted as the site was close to commercial shipping lanes) and Granite Peak in Utah, a substation close to Dugway Proving Grounds (which will make a re-appearance in Chapter 6). The Vigo Production facility was added in Indiana. Vigo was designed to produce biological agents and ammunitions on an industrial scale. In addition to this extensive complex of scientific centres, WRS also established a widespread research network with US universities. At its height the programme employed almost 4,000 workers and cost $60 million during the war (Bernstein 1987: 116).

President Franklin D. Roosevelt

It is uncertain what Roosevelt actually knew about this programme. Roosevelt verbally approved the WBC report two weeks after receiving Stimson's memorandum, which suggests he was aware that a programme was in the works. Some analysts claim, however, that the details were kept secret from him (Bernstein 1987: 116–18; Guillemin 2005: 60). It would not be the first time that a national leader had been left in the dark on biowarfare—and explicitly for reasons connected to the taboo. Japan's Emperor did not know that a state programme existed. He was not told in order to protect him from any negative association with a reprehensible mode of aggression in line with the taboo. Ishii (in S. Harris 2002: 1) himself said: 'Biological warfare is inhumane and advocating such a method of warfare would defile the virtue and benevolence of the Emperor.' The Emperor could not be seen to accept what was intrinsically unacceptable. While bioweapons were pursued by Japan, therefore, this development happened only under the belief that a clear and eminent norm was being broken in the process. Moreover, the norm's breaking was interpreted as a fundamental (if not entirely prohibitive) wrong.

Roosevelt had repeatedly expressed his opposition to chemical warfare. He emphatically rejected chemical weapons as immoral and abhorrent. He said: 'It has been and is the policy of this Government to do everything in its power to outlaw the use of chemicals in warfare. Such use is inhuman and contrary to what modern civilization should stand for' (Roosevelt 1947). A SIPRI (1971b: 316) report asserts that the President's 'hatred of gas' was so intense as to explain any lack of US preparedness for chemical warfare during World War II. If the US did not wage chemical violence during the conflict it was not because it was incapable, per a practical measure of capacity. Unpreparedness existed because Roosevelt did not *want* to be capable, per the taboo. Roosevelt's revulsion was further evident in his refusal to approve a CWS request to change their title to the Chemical Corps (CC). He believed that the move would legitimize chemical warfare at a time when he sought to eradicate it. In a letter to the Senate opposing the change he said:

> I am doing everything in my power to discourage the use of gases and other chemicals in any war between nations. ... I do not want the Government of the United States to do anything to aggrandize or make permanent any special bureau of the Army or the Navy engaged in these studies. I hope the time will come when the Chemical Warfare Service can be entirely abolished. To dignify this Service by calling it the 'Chemical Corps' is, in my judgment, contrary to a sound public policy.
>
> **(Roosevelt 1947)**

Roosevelt also endorsed a defensive policy of no-first-use pertaining to chemical warfare:

> From time to time since the present war began there have been reports that one or more of the Axis powers were seriously contemplating use of poisonous and noxious gases or other inhumane devices of warfare. I have been loathe to believe that any nation, even our present enemies, could or would be willing to loose upon mankind such terrible and inhumane weapons. ... Use of such weapons has been outlawed by the general opinion of civilized mankind. ... I state categorically that we shall under no circumstances resort to the use of such weapons unless they are first used by our enemies.
>
> **(Roosevelt 1943: 507)**

Consequently, Roosevelt's public commitments were expressed in the very deepest terms of taboo. The President's core themes of poison, inhumanity, and upmost loathing in the above quote all demonstrate this point.

Roosevelt never made a public statement against biowarfare. It is accepted, however, that his repulsion towards chemical violence was similarly applicable to the biological (USCong 1976: 360; see also Geissler, van Courtland Moon, and Pearson 1999: 257; Tucker 2002: 109–10; Guillemin 2005: 9). Officials may then have limited what the President was told about the bioweapon programme fearing that he would be appalled and attempt to stifle it. Yet Roosevelt was approached concerning hypothetical proposals to employ biowarfare tactics. Roosevelt's Chief of Staff and close friend, Admiral William Leahy (1950), recounted a discussion concerning bioweapons in July 1944:

> I recall particularly that as we were sailing for Honolulu for the MacArthur-Nimitz conferences in July of 1944, there was a spirited discussion of bacteriological warfare in the President's cabin. By that time scientists thought, for example, that they could destroy completely the rice crop of Japan. Some of those present advocated such measures. Personally, I recoiled from the idea and said to Roosevelt, 'Mr. President, this [using a bioweapon], would violate every Christian ethic I have ever heard of and all of the known laws of war. It would be an attack on the noncombatant population of the enemy.'
>
> **(Leahy 1950: 439–40)**

Leahy did note that the President remained 'non-committal' to the suggestion of biowarfare. The proposal, however, was ultimately not approved nor supported. Moreover, the proposal can be further contextualized in that it referred only to crops. Targeting plant life may be viewed as incompatible with the highest ideals of the taboo—as Leahy's reaction demonstrates. Yet the offensive biological plot was explicitly not aimed at humans. In the same way that the targeting of animals during World War I (in Chapter 3) can be viewed as a highly qualified notion of bioweapon use that may not necessarily violate taboo thinking, so the emphasis on crops here can also be seen to represent a conditional understanding of biowarfare that still recognized the moral limitations imposed by the taboo. As such, the influence of the taboo was still significant in respect to Roosevelt and his administration.

President Harry Truman

President Harry Truman (1953) wrote on the day before his presidential term of office ended that he would have resorted to biowarfare if World War II had not been over by 1945. Harris (2002: 212–13) said that, while Truman 'left no paper trail as to his attitude toward BW [bioweapons]', he would have done anything to win. Harris continues: 'Truman ... was a simple but strident patriot. He believed that when the country is in a life-and-death conflict, the armed forces should use any weapon available to defend American freedom.' This view, however, is reductionist. Truman did *not* authorize or use bioweapons. Biowarfare was still constructed as (at most) the means of last resort in that Roosevelt's no-first-use posture remained in place at this time. Moreover, there is no other evidence that Truman would have actually pushed the biological button despite his hard-line, but still speculative, declaration. Barton Bernstein (1987: 121) points out: 'Germ warfare, with its specters of epidemic and invisible poison, might have been harder to endorse.' Truman may have discovered that biological attack was not the viable option that it appeared in theory thanks to the taboo. Whatever anyone believes they *could* do, the taboo may yet preclude action when faced with the actual decision to do so. Hypothetical claims that an actor is capable of inflicting mass death through immoral means, and is also willing to brave the subsequent backlash that would accompany such behaviour, should be taken with a pinch of salt.

Harris' argument is additionally problematic in that it is predicated on Truman's engagement of nuclear arms against Japan. Harris (2002: 213) claims that a man who was prepared to use nuclear weapons was surely prepared to use biological ones also. Equating biological and nuclear weapons in this case, however, is untenable. Tannenwald (2007: 73–114) has established that there was no clear taboo pertaining to nuclear armaments until *after* these weapons were employed. Prior to the use of nuclear weapons at Hiroshima and Nagasaki the 'assumption was that atomic weapons were legitimate weapons of war' (Tannenwald (2007: 79). Tannenwald's claim is reflected in Truman's (1955: 462–3) own account of authorizing the attack on Japan. Truman said: 'Let there be no mistake about it. I regarded the [nuclear] bomb as a military weapon and never had any doubt it should be used.' In contrast, the previous chapter has outlined a longstanding taboo pertaining to biowarfare and the Geneva Protocol was already 20 years old when Little Boy was detonated. Consequently, Truman's approach to a tabooed weapon (biological) is neither identical nor comparable to one that was not clearly considered pariah at the time of its first employment (nuclear).

Nuclear weapon use against Japan, therefore, is not sufficient evidence that Truman was prepared to violate the bioweapons taboo. Indeed, Truman (1948) actively spoke out against bioweapons after the war. He expressed himself in terms that reflected the taboo and the horrific levels of destruction that these weapons could perpetrate. In an address to the American Legion Convention, the President said:

> The horrors that modern war inflicts on innocent people are known to all Americans. The use of atomic weapons and bacteriological warfare, in particular, might unleash new forces of destruction which would spare no nation. All of us are well acquainted with that fact.
>
> **(Truman 1948)**

While the identified statement was obviously made within the luxury of peacetime, Truman still openly expressed his strong opposition to biowarfare. The taboo was a factor in his political thinking.

An observation by Paul DeForest summarizes the wider situation extremely well. He notes that the biowarfare programme took place '*despite*' [emphasis in original] international normative commitments against these weapons (DeForest 1990: 24). Yet he adds that 'in comparison to the systematic violation of the norm protecting civilians by strategic bombing, the record of adherence to the prohibition against BW [bioweapons] ... was much better during World War II'. In a situation where the extreme pressures of a global war had undermined all the social and political expectations of conflict, the bioweapons taboo was relatively less strained than other norms. The only clear allegation of biowarfare during World War II relates to the 1942 assassination of German SS official Reinhard Heydrich as part of Operation Anthropoid (SIPRI 1971b: 110; R. Harris and Paxman 2002: 70–108; Tatu, Jost, and Bogousslavsky 2017). Allied agents attacked Heydrich's car as it travelled through Prague, stopping the vehicle with an anti-tank hand grenade and then opening fire with a machine gun. It is claimed that either the grenade or bullets contained botulinum toxin. The allegations, however, are unconfirmed—and World War II concluded without proven use of bioweapons.

The backlash against biowarfare

The end of World War II was almost the end of the US biowarfare programme. On 11 August 1945 (three days prior to Japan's surrender) the War Department informed the CWS that the biowarfare programme was to be transferred to peacetime status; in effect, it was shut down. With America's bioweapon development strategy seemingly over, there would soon be recrimination that it had ever existed. While many political and military actors may have consented to temporarily put their ethical scruples aside during World War II for what was perceived to be the greater good, the US public would be less forgiving.

The backlash began with the publication of the Merck Report in January 1946 (Merck 1946). George Merck, President and Chairperson of the Merck & Co pharmaceutical company, had headed up the WRS during World War II. He was not the first choice for the role. Three other men had rejected the position before him: Chair of the Rockefeller Foundation, Walter Stewart; President of John Hopkins University, Isaiah Bowman; and President of Cornell University, Edmund Ezra Day. The extensive list of unwilling candidates is perhaps a sign that the job was itself taboo due to its biologically aggressive associations. Merck's report was an account of the programme that he had overseen, and its publication was the War Department's attempt to effectively come clean. Seeking to vindicate the programme, the report's content argued that biowarfare *did* pose a major threat to US national security. The report also presented biowarfare as a viable method of military aggression, despite a lack of evidence that the wartime programme had produced workable armaments and despite Merck's own admission that there remained no consensus on the utility of bioweapons. It was felt that biowarfare needed to be highlighted as a credible threat in order to then justify why the US would require a bioviolence capacity in response.

Merck's justification for the US programme also reflected the taboo. Merck (1946: 237) maintained the line that offensive research was pursued for defensive means only. Another of the report's authors and secretary to the committee that produced it, R. Gordon Arneson (1989: 50), commented later that 'nobody had in mind that we would have any use of bacteriological weapons in war, but it was something that ought to be looked at in the future'.[5] Consequently, biowarfare was again constructed as a highly conditional concept that rejected actual use. Merck additionally framed the programme in terms of its supposed humanitarian benefits and repeated the moral sentiments expressed within taboo thinking. The study highlighted the supposedly near-philanthropic 'advancement of knowledge' generated by the programme in relation to agriculture, medicine, and 'human welfare' (Merck 1946: 240–1). To the extent that Merck's argument referred to medicinal benefits in particular, this construction again attempted to associate biowarfare positively with public health in order to undermine negative perceptions (per Stimson's emplacement of WRS within the FAS remit). The authors then explicitly drew upon the taboo to sell the programme in its best possible light

[5] Arneson (1989: 49–50) additionally said of writing the report: 'It was prepared, and it was issued, and then withdrawn from public service. Nobody knows where it is. I used to have a copy of the darn thing; I lost it years ago.'

as part of a strategic narrative designed to make biowarfare palatable to a public audience.

The approach did not work. The public were horrified.[6] An outcry of anti-biowarfare sentiment followed throughout the media. The *New York Times*' Military Editor, Hanson W. Baldwin (1946a: 52), vilified and labelled bioweapons as 'The Black Death'. He said: 'Disease is a weapon of surprise and stealth; it can attack suddenly and swiftly or with slow deliberation.' Baldwin (1946b: 4) elsewhere discussed the importance of the Geneva Protocol as a critical 'legal and moral barrier' in preventing the insupportable terror of bioviolence. Other articles within the newspaper compared bioweapons to the 'horror' of the atomic bomb (Kaempffert 1946: 14):

> In a sense we have something far worse than the atomic bomb. After all, Hiroshima and Nagasaki were wiped out in an instant and the after-effects of radioactivity proved to be less alarming than was at first supposed. But infectious diseases? They spread like slow fires; they cripple a nation if not immediately checked by drastic measures; they delay indefinitely the reconstruction of a devastated country; they entail the maintenance of a quarantine for years; they are two-edged swords against which attackers and attacked must defend themselves.
>
> **('Pandora's Box', 1946: 14)**

Lurid tales were told of the appalling things that bioweapons could supposedly do. Editorial Director of *Science Illustrated* magazine, Dr Gerald Wendt, made headlines after he announced that the US military had access to a crystalline toxin so potent that only an inch-sized cube of it could kill everyone in the US and Canada ('Ounce of New Poison', 1946: 1). Wendt disparaged bioweapons as 'silent, insidious, anti-human weapons that make no sound, give no warning, destroy no forts or ships or cities, but can wipe out human beings by the millions'. *Life* magazine infamously published a highly graphic article with a shockingly vivid photograph of a biological researcher donned in protective clothing and carrying an oxygen mask. The article debated the issue in openly moral terms. Biowarfare was part of the 'equation of good and evil' in which 'frontiers of the public domain have been staked out in the name of military secrecy' (Piel 1946: 130).

Highly visible public opponents spoke out against bioweapons. Such figures are termed here 'taboo advocates' in that they drew upon taboo arguments to make their case. Chisholm described nuclear warfare as a 'has-been'

[6] Public response was somewhat conditional. A day after the Merck Report was released, the Navy announced at a press conference that it had infected 50 prison inmates from San Quentin with bubonic plague at its Naval Research Unit at Berkeley, California ('Convicts Aid Germ Battle', 1946). There was no public backlash to this revelation. Regis (1999: 94) notes in mitigation, however: 'There was no public uproar then as there might have been in later times: the men, after all, had volunteered for the project . . .'

in comparison to the terrors of germ conflict as the latter threatened the very survival of the human race ('Bacteria Put Above A-Bomb', 1947: 20). The atomic bomb was 'child's play compared to biological weapons' ('Atomic Bomb', 1949: 7). Theodor Rosebury was another notable public activist. Rosebury was a scientist who had headed up the biowarfare programme at Camp Detrick. He had initially seen a defensive need for bioweapons as an unwelcome but obligatory means during World War II. As such, he had been prepared to temporarily put his moral reservations aside in respect of his patriotic duty:[7]

> We were fighting a fire, and it seemed necessary to risk getting dirty as well as burnt. … We resolved the ethical question just as other equally good men resolved the same question at Oak Ridge and Hanford and Chicago and Los Alamos.
>
> **(Rosebury 1963: 514–15)**

The reference to biowarfare as 'dirty' again comprises a telling representation of the taboo in Rosebury's thinking. The scientist also understood biowarfare as inherently unethical and his perception of bioviolence in taboo terms would prove overwhelming. Rosebury would quickly remove himself from the programme because of the immorality and pariah status that he associated with these weapons. He was revolted by what he had done at Detrick (Clarke 1968: 19). Rosebury sent a private memo to his superiors in 1946 expressing horror at, as Hersh (1968a: 300) describes it, 'what kind of monster he and his fellow researchers at Detrick had created'.

Rosebury became a key campaigner in the public fight against bioweapons. He published a famous article in 1947 with fellow scientist Elvin Kabat in *Journal of Immunology* undermining biowarfare and calling for peace. A newspaper article said of the publication: 'The authors made it plain when they released the report to the press that they were more interested in abolishing war than in making it more terrible' (Kaempffert 1947: 11). Rosebury expanded on the work in a book: *Peace or Pestilence: Biological Warfare and How to Avoid It* (which has more recently found fame by being included on Al Qaeda reading lists). Here Rosebury (1949) expressed his repugnance towards biowarfare in repulsed and taboo-like terms:

> The power of destruction is now so great on both sides that, once we let loose in the inexorable chain reaction of war, the clock of civilization may be turned

[7] Brian Balmer (2002b) identifies similar motivations with respect to scientists working on the British bioweapon programme at this time.

> back centuries, if not millennia. Perhaps we will leave the world to the rats and the cockroaches.
>
> (Rosebury 1949: 3)

Rosebury leaves his reader with a picture of bioweapons as inimical to civilized behaviour and associated with the most disgusting of creatures.

Other scientists similarly felt that the very nature of biowarfare conflicted with the moral ideals of their profession. They believed in saving lives; not finding ever more efficient means to take them by the same method (as Simmons had understood in relation to the Hippocratic Oath). It is no coincidence that the Army Medical Corps initially refused to station a medical team at Camp Detrick (Langer 1968: 120). Opposition galvanized within the wider scientific community following the Merck Report. The International Congress for Microbiology (ICM) adopted a resolution in 1947 calling for an outright ban on bioweapons. The statement 'considered such barbaric methods [biowarfare] as absolutely unworthy of any civilized community' (Laurence 1947: 12). The ICM participants (including representatives from the US) said that their aim of prohibiting bioweapons would be achieved by:

> [M]oral persuasion, telling the world of the possible terrors of bacteriological warfare and thus arousing the conscience of the world's peoples to a degree where civilized nations will come to realize the urgency of the need for a universal agreement outlawing such methods.
>
> (Laurence 1947: 12)

The ICM called upon the morality and notions of civilized conduct intrinsic to the taboo to make its case. The concern was not understood as a question of military or medical necessity; biowarfare was explicitly an immoral issue that must be addressed and challenged on normative terms.

Overall then, a varied and extensive groundswell of public opposition had emerged against the programme. Even Merck himself would call for the abolition of bioweapons the following year ('Merck Urges World', 1947: 2). The post-war backlash, therefore, supports Tannenwald's (2007: 47) assertion of a taboo as a society-up construction that relies heavily on domestic public opinion. Wider public resistance based on taboo sentiment comprised a key factor constraining biowarfare policy. Critically, this response was not just a resistance to their use, but also to the very possession of biological armaments.

From World War to Cold War

If the development of a US biowarfare programme can be explained in relation to the pressures of World War II, then its second life can be credited to a further conflict: the Cold War. The Communists would seemingly come to comprise a greater danger than domestic disapproval. Biowarfare became intrinsically connected to the Soviet risk inherent in the Cold War and the need to retaliate against it (CIA 1947). It was claimed within military and intelligence sectors that Soviet biowarfare capabilities would 'progressively increase' to the point at which the Union of Soviet Socialist Republics (USSR) would 'probably be capable of serious air attacks against the United States' (Souers 1948: 665; see also Steelman 1949: 344) as well as sabotage strikes ('National Intelligence Estimate' 1951: 1624). Consequently, American interest in biowarfare increased as key actors argued that the US needed to prepare for a Soviet biological assault.

There was very little evidence that the Soviet enemy possessed workable or extensive bioweapons (CIA 1948; see also CIA 1949a, 1949b). CIA Director, Rear Admiral Roscoe Hillenkoetter (1948), admitted that there was 'only meager information' concerning the USSR biowarfare capacity and that significantly more intelligence was required to make an informed assessment (see also Machle 1949: 1012–13). Intelligence gathering at the time was highly limited and dependent on unsophisticated tactics such as monitoring propaganda in the foreign press. Hillenkoetter (1948) recommended finding 'a dissatisfied Soviet scientist, intimately connected with the program, who is willing to disclose it'—interestingly, a statement that clearly assumed there was a significant programme to uncover.[8] Hillenkoetter's view, however, was not universal. Other US officials did not believe that a major Soviet bioviolence threat existed. It was stated at a 1949 Policy Planning Staff meeting on nuclear conflict that Russia simply could not finance all types of armament and so 'must choose her weapons for war' ('Minutes of the Eighth Meeting', 1949: 204). It was then further reasoned that the Soviets would not prioritize biological means of aggression as part of such a calculation. The minutes concluded: 'Biological warfare, which at one time seemed to be a real menace, does not now seem likely to be of major importance in case of war' ('Minutes of the Eighth Meeting', 1949: 204).

[8] Hillenkoetter (1948) also proposed the following plan: '[W]e should supply Russia, at prices sufficiently low so that internal production was impractical, such biological preparations as serums, vaccines, penicillin, etc. This might deter them from the construction of productive plans for such materials. Such plants would be easily converted to biological warfare purposes ....' That is, sell the Soviets cheap drugs so that they would not build the facilities that could then be used for making bioweapons.

Yet the opinion that Russian bioweapons *did* pose a meaningful threat persisted elsewhere. Van Courtland Moon (2009: 9) explains this view: 'The Cold War conditioned the US to see the USSR as a formidable and determined enemy whose evil intentions would lead to development of every possible weapon to secure world domination.' He continues: '[E]vil governments will do evil things; therefore, the US [felt it] must be prepared to retaliate in kind because it must be ready to meet every possible contingency so as to punish the evildoers with their own poison' (van Courtland Moon 2009: 43). Robert Harris and Jeremy Paxman (2002: 142) make a further claim. The authors argue that, while the development of nuclear arms meant there were those within the US 'who argued that the distasteful business of waging war with disease could be forgotten', this was not deemed to be the case in respect of the USSR. The Soviets did not yet possess nuclear weapons—and would not until 1949. Consequently, it was posited that the USSR would find bioviolence attractive as a means to offset both US nuclear prowess and their own lack of nuclear capacity. Concerns surrounding a Soviet biowarfare capacity continued, however, even once Russia acquired nuclear arms. Oliver Edmund Clubb, Director of the Office of Chinese Affairs, sent a memorandum in November 1950 to then Assistant Secretary of State for Far Eastern Affairs, Dean Rusk. Clubb (1950: 1091) warned that the Soviets may be planning to use 'bacteriological warfare, which it estimates will balance the atomic bomb'.

Cold War fears were also manifest within America's decision to grant secret immunity to those involved in the Japanese wartime programme. Supreme Commander of the Allied Forces, General Douglas McArthur (e.g. Harris 2002; Guillemin 2005: 75–91, 2017), engineered an offer to excuse the Japanese scientists including Ishii. McArthur further promised that the US would not pass on information concerning the Japan programme to the War Crimes Tribunal. America obtained Japanese data on human testing in return. While the US may have challenged the bioweapons taboo in developing a programme (one that involved limited trials on people), the state had yet to break the taboo surrounding lethal human biological experimentation as carried out by the Japanese. Japan's programme, therefore, provided US scientists with biowarfare data that they would not otherwise have access to or be able to create themselves. Significantly, however, US policymakers did still uphold the taboo in respect of other state programmes. The Nazi biowarfare programme, for example, was publicly condemned as intrinsically immoral (Brody et al. 2014).

Strong bureaucratic forces were also at play. Military actors including the CWS had become heavily invested in a US biowarfare programme during World War II to the extent that this comprised a point of institutional survival and esteem. Wartime activity concerning bioweapons had created a significant regime around which there was momentum to continue. Abandoning the programme would effectively ask the CWS and others to cede the substantial power and influence that the previous push towards a biowarfare capacity had facilitated. US Congressperson Richard McCarthy (D-New York; 1970: 123) later highlighted this influence of the CWS as a key factor in domestic biowarfare policy. He commented that the CWS had asserted 'slow but constant pressures' on the US government in order to gain support for the programme. Explicitly, McCarthy (1970: 124) discussed this influence as an intrinsically bureaucratic one in which the key figures involved had been working 'with the development of this technology throughout their lives'. With such credibility, power, and resources at stake, this institutional thrust combined with fears of a Soviet attack and new data from the Japanese to keep a US programme in contention. Even for those who did not accept that a USSR biological capacity comprised a foremost concern, that prospective threat was an ideal excuse for actors within the biowarfare regime who wanted to keep their job.

Yet the taboo was still a factor. Clark Clifford (1971), Special Counsel to the President from 1946 to 1950, sent a memorandum to the President at this time outlining areas of Soviet vulnerability that could be subject to military exploitation. Clifford identified biowarfare as one such area and concluded that 'in order to maintain our strength at a level which will be effective in restraining the Soviet Union, the United States must be prepared to wage atomic and biological warfare'. Speaking about his memo in 1971, however, Clifford qualified his apparently bellicose statement:

> I was indicating that we should be prepared to undertake the defense in any way that became necessary. ... As far as biological warfare is concerned, it would be perfectly alright with me if we destroyed every form of biological means of warfare that we have in this country.
>
> **(Clifford 1971)**

Consequently, the push towards biological aggression at this time was not devoid of normative influence. Clifford expressed a deeply taboo abhorrence of biowarfare despite his concerns surrounding the Soviet threat. Nor could the drive to reinvigorate the programme ignore continued public and political opposition to biowarfare as a taboo concern. For example, the clandestine nature of the Japan deal was not simply a desire to preserve military secrets. Immunity comprised a case wherein the deal's orchestrators were concerned about the taboo association with repugnant forms of warfare and that the 'grotesque details of the Japanese programme could well have tainted the Camp Detrick program' (Guillemin 2005: 88). The pervading secrecy surrounding the biowarfare research deal with Japan was, therefore, a direct reaction to the taboo.

The taboo also shaped the public presentation of biowarfare policy. The CWS's name was now amended to the Chemical Corps. Yet the move was not the aggressive or bellicose act that Roosevelt had feared—but was now a means of portraying the organization as 'less warlike' and 'considerably more innocuous' (Regis 1999: 116). Army Chief of Staff, Dwight Eisenhower, also banned the military from all public mention of the term 'biological warfare' in that the issue was recognized as highly contentious (Brungs 1964: 69). Similarly, the Merck Report was now classified and not released again until 1974 due to the controversy that it had engendered. The examples here may not indicate that military opinion had changed as to the rights and wrongs of biowarfare. The clear shift in presentation, however, comprised a reply to the pressure of the taboo as vented through public and political channels. Officials acknowledged that wider national and international audiences expected the US to adhere to certain normative expectations and that their actions had to be seen to conform—whatever else was discussed behind the scenes in promoting a biowarfare capacity.

The taboo was further evident in a major statement in 1949 by Secretary of Defense, James Forrestal (1949: 37). Forrestal's position was fundamentally opposed to the taboo. His statement was interpreted as breaking the reticence towards even mentioning biowarfare in order that he could promote these weapons in a positive light. Forrestal 'broke an official three-year military silence' on the issue of bioweapons and 'ended what had in effect been a blanket taboo on the utterance of the phrase "biological warfare"' (Waggoner 1949: 1). The Secretary also condemned the negative media response to the Merck Report, which he described as 'so sensational as to be completely misleading' (Forrestal 1949: 37). His statement was, in essence, a call to educate those public and political actors who had rejected bioweapons that these armaments were actually a necessary and valid form of aggression. Those repulsed by these weapons must be set straight. In effect, biowarfare was seen to have attracted a stigmatized label of wrongdoing and this label must now be reversed.

Yet Forrestal's speech was framed in respect of the taboo—albeit to justify a 'non-taboo' position. With strong echoes of Merck, Forrestal highlighted the supposedly limited destruction of bioweapons so as to neutralize the taboo idea that bioweapons were devastating armaments. His argument also focused heavily on biowarfare as a defensive construct. He promoted a more conditional and palatable understanding of bioviolence by emphasizing the civilian and non-offensive aspects of the programme.[9] Forrestal further played up the public health and medical benefits of biowarfare research. Consequently, his statement recognized that: there was a taboo on bioweapons; the taboo was sufficiently influential to shape policy; and arguments in favour of biowarfare must be constructed in direct in response to taboo expectations in order to be considered legitimate. Forrestal may not have personally approved of the terms of the taboo. Yet he also accepted that the taboo was so strongly felt elsewhere that the norm comprised the position to which he needed to directly respond so as to promote his own stance.

Educating against the taboo

Forrestal's call for public education on biowarfare was taken up by the Stevenson Committee. The Committee was established at the direction of Secretary of Defense, Louis Johnson, in 1950 to investigate the viability of the US bioweapon programme. Earl Stevenson chaired the committee. Stevenson was then president of the management consultancy firm, Arthur D. Little. The company was already one of the US government's own specialist consultants in developing biowarfare systems and it was the recipient of a number of lucrative military contracts.[10] Consequently, as with Merck, Stevenson had significant commercial interests that favoured the continuation of the programme. Furthermore, the committee carried out its proceedings at the same time as the Korean conflict was emerging. The final report was released less than a week after the war broke out. It was unsurprising then that the committee's findings were highly supportive of bioweapons and the US need for them.

The Stevenson Committee proposed a number of aggressive recommendations in relation to developing the US biowarfare capacity (Stevenson 1950). The final report called for: a commitment that bioweapons would not be subject to a retaliation-only policy; all 'necessary steps' to be taken to develop a workable biowarfare programme; the finalization of all plans for future research and facilities development; comprehensive scientific tests to decisively determine the military value of bioweapons; a significant and material expansion of the programme; and a public information campaign to educate

[9] CWS Chief, Major General Alden Waitt, made a comparable statement at this time. Waitt (in Waggoner 1949: 1) asserted that, while the US needed to be prepared for enemy use: 'It does not follow that if the enemy uses gas or biological warfare that the United States will use them.'

[10] The company's laboratory famously manufactured a literal silk purse made out of sows' ears in 1921. The purse is now held by the Smithsonian Institute.

on the positive value of biowarfare (Stevenson 1950: 1–2). On the last point, the committee argued that a campaign was necessary in order to target the 'contradictory, exaggerated, distorted, and highly sensational' news reports that the public had been subject to (Stevenson 1950: 21). Secretary Johnson (in Stevens 1950: 78) similarly said that improved public understanding of these weapons would 'control fear with reason instead of exaggerating it into hysteria'. The US government accepted all the Stevenson proposals bar one: officials refused to abandon a retaliation-only position.

The Stevenson Committee proceedings, however, still acknowledged that the bioweapons taboo comprised an important and influential factor. Brigadier William Creasy was then the head of the CC's Research and Engineering Command and later CC Chief. Creasy gave testimony to the committee in which he accepted that there *was* a psychological element to biowarfare of the type connected to the taboo. Bioweapons created a special type of fear. While Creasy (1950) spoke of this fear as being evident in an 'uninformed public' who did not know any better, he acknowledged:

> There is no doubt that bacteriological warfare offers a unique psychological hazard. ... Man's dread of disease is universal. The mysteriousness and invisibility of bacteriological warfare agents, the knowledge that they strike via the simplest and most basic sources of man's security—food, drink, and the air he breathes—and a feeling of helplessness in dealing with the unknown, all add to the psychological potential. A combination of psychological effects and reduced rations would be expected to diminish a nation's will and ability to fight.
>
> **(Creasy 1950: 21)**[11]

Creasy's statement, therefore, recognized the psychological perceptions that exist around biowarfare and which are associated with a taboo feeling. Even more interestingly, Creasy recognized that this taboo had the potential to exert real effect in terms of these weapons' use. Bioweapons had a psychological impact that was connected to battlefield efficacy. Consequently, even the most hard-line elements within the US programme accepted the presence and significance of the taboo, even if they did not seek to ascribe to its normative rule themselves.

The Stevenson Committee attempted to negate the taboo. The norm's repudiation was expressed through an emotive appeal to military necessity and a

[11] Creasy (in 'Enemy Disease Smog', 1951: 18) said elsewhere: 'Biological warfare is essentially public health and preventive medicine in reverse'. Consequently, Creasy repeated and openly expressed ideas that are intrinsically connected to the taboo and its negative associations of repulsion, especially the idea that bioweapons are antithetical to health.

suggestion that moral arguments did not exist with respect to bioweapons (Stevenson 1950: 9). The committee put forward the following historical rationale: the American Revolution had modified and evolved the needs of war. Prior to the Revolution, wars had been 'professional'. The Revolution, however, had pushed conflict towards a type of total war: 'In order to conquer a country it became necessary to conquer a people. To defeat a professional army was not enough. This led to all-out wars' (Stevenson 1950: 9). Consequently, the morals that had previously characterized the conduct of war were simply no longer applicable regardless of any desire to retain them. Moreover, the committee argued that any attempt to cling to such idealistic values would only give advantage to the enemy. Committee members further contended that biowarfare could not (as charged) increase 'the horrors of war' in that this had already been achieved with other weapons (Stevenson 1950: 10). Even conventional arms could be employed to extreme destructive effect. If all war was now total in terms of actor motivation, therefore, then it no longer mattered precisely how that totality was achieved. The committee effectively claimed that biowarfare did not comprise a negative moral conundrum; bioweapons were not taboo. The very fact that the Committee felt they had to demonstrate that bioweapons were not taboo, however, says much about that norm's perceived strength. Why feel that you need to explain away the taboo if you do not accept that it is a valid concept, at least to your audience?

Indeed, the subsequent public relations campaign in favour of bioweapons was constructed within the ideological framework of the taboo. For example, the Stevenson Committee stated that a campaign should focus on bioweapons development as a purely defensive concept (Stevenson 1950: 21–2). While the restriction of policy to defence was explicitly *not* the opinion of committee members (in that they had openly lobbied for the elaboration of offensive options), it was recognized that offensive policies were considered pariah. A clear acceptance existed that bioweapons were not condoned more widely and that these armaments were stigmatized within the context of the taboo. Notably, the committee's stance yet again supports arguments that the taboo explains state secrecy surrounding biowarfare. The desire to keep offensive action covert in this case was explicitly not a consideration of national security but a direct response to the potentially negative backlash that would ensue as a consequence of taboo feeling.

In a more specific example, the Federal Civil Defense Administration published a 10-cent brochure entitled 'What You Should Know About Biological Warfare' (P. Kennedy 1951: 31). The booklet sought to challenge the negative public reaction that had dogged plans for an expanded programme. The content reflected both Forrestal's comments and the Stevenson recommendations by directly engaging with taboo perceptions of biowarfare. The booklet stated that the mass destructive potential of biowarfare was a 'myth' and that 'talk of poisons that can kill millions is nonsense' (Kennedy 1951: 31). The booklet also listed six core 'survival secrets' pertaining to biowarfare. Number six cautioned: 'Do not start rumors—do not believe wild stories. Start a rumor and you may start a panic.' The statement implied that any concerns about biowarfare were simply scare stories and that communicating those fears publicly was little better than spreading propaganda. Consequently, the publication was designed to undermine the taboo by effectively reducing it to a conspiracy theory. Yet again, however, the attack on the taboo was also evidence of the norm's significance. If the taboo was of no import, why then be concerned with denigrating it? The concept of the taboo was held up by some as an ideal to be trounced. The motivation inherent to that desired rejection, however, necessarily accepted that the taboo was sufficiently powerful as to constitute a major barrier to a biowarfare programme. The sheer effort applied in undermining the taboo through public campaigning demonstrates normative robustness and the extensive impact that actors perceived the taboo to have on security politics at the time.

Bioweapons as weapons of mass destruction

The Stevenson Committee was also compelled to respond to another aspect of the taboo: the emerging international debate on eliminating WMD. The newly created UN had responded to the extreme demonstration of nuclear force at Hiroshima by seeking to control not only nuclear armaments but all weapons capable of similar levels of devastation. The WMD concept was introduced in a November 1945 communiqué issued by Truman together with British Prime Minister Clement Atlee, and Canadian Prime Minister McKenzie King. The statement called for a new commission to eliminate 'atomic weapons and of all other major weapons adaptable to mass destruction' ('United States—Great Britain—Canada: Declaration on Atomic Energy', 1945: 50). A later definition would identify these other weapons as biological, chemical, and radiological armaments (UN 1948).[12]

[12] The UN's exact wording was: '[WMD are] atomic explosive weapons, radioactive material weapons, lethal chemical and biological weapons, and any weapons developed in the future which have characteristics comparable in destructive effect to those of the atomic bomb or other weapons mentioned above.'

Far from accepting biowarfare, therefore, Truman (the man who had claimed that he would have employed bioweapons in order to end World War II) was actively advocating for its condemnation and abolition as one of the international community's highest priorities. It was initially uncertain that the WMD concept would include biological arms. Yet the document's author, Vannevar Bush—an American engineer and scientific advisor who had headed up the Office of Scientific Research and Development (OSRD) agency throughout World War II[13]—was determined to include bioviolence in the definition. Bush (1970: 298) commented: '[W]hile we were attempting to bring reason to bear on one terrible weapon [nuclear], we might as well include another [biological] that could be equally as terrible.' Truman and Bush's inputs demonstrate that US actors were leading the way in bringing about a global cessation to biowarfare. American influence was central to both the stigmatizing and fetishizing processes that underpinned the growing taboo.

Admittedly, the WMD concept was not a straightforward political commitment to disarmament. The proposed ban on biowarfare was (at least in part) a diplomatic project that was strategically shaped and manipulated in line with wider Cold War politics (Bentley 2014: 32–48). The very concept of WMD was manifest as a series of political moves in which foreign policy actors sought to control the arsenals of others. Cherry-picking weapons to prohibit under the banner of WMD was a decision concerning (1) which armaments certain actors wished to see removed from their opponent's military stockpiles and (2) what weapons they were prepared to sacrifice from their own possession in order to achieve this. Furthermore, concerns surrounding WMD were also motivated by more self-interested considerations. Eisenhower (1946) commented that a nuclear weapons ban must be accompanied by prohibitions on comparable weapons in that these other armaments could potentially take the place of an atomic capability. The WMD concept, then, was not purely a political construction focused on achieving disarmament but also a pragmatic means of covering all bases from a practical military perspective. These other factors, however, should not detract from the normative commitment to eliminating pariah weapons driving the diplomatic push for major disarmament. The UN agenda still reflected the taboo.

The UN conceptualization of WMD constituted a high-profile restatement of the taboo. Consequently, the term was also a threat to any US biowarfare programme. The Stevenson Committee sought to nullify the challenge of

[13] OSRD was responsible for wartime military research including bioweapons development and the Manhattan Project.

the taboo by questioning the WMD concept and asserting that bioweapons had been 'erroneously' included (Stevenson 1950: 8). Committee members argued that it was unclear whether bioweapons were capable of mass destruction and would so warrant the controversial label. The claim was less a point of procedural or definitional clarification as a further attempt to destabilize the taboo and any ideological rejection of a US biowarfare capability. The committee's questioning of the taboo did not, however, undermine the wider international commitment to bioweapons' elimination. The call for disarmament had set in motion a clear and highly public process of stigmatization and fetishization that enjoyed the institutional backing of the UN as well as wider international support. Furthermore, the condemnation of biowarfare was not simply a statement of moral wrong. UN activity on the matter sought to actively codify this sentiment in real diplomatic commitment as a priority. In the same way that the Geneva Protocol had stigmatized the taboo by crediting it with an official and customary status, so the UN discussions gave formal political weight to the normative ideas opposing biowarfare and established the threat as a recognized staple of international debate. Political actors had made a clear commitment to the idea that bioweapons were taboo and ought to be eradicated as a consequence. Indeed, the multilateral rejection of biowarfare as an immoral and pariah act was now the very poster child of the UN mission.

No bomb ≠ no taboo

At the end of the day there was no major biological bomb. When the Stevenson Committee sat to discuss the future of the bioweapons programme, that programme had yet to mass produce a workable device suitable for the battlefield. The absence of an effective bioweapon opens up a potential readiness claim that non-use was not a consequence of a taboo. The failure to employ a bioweapon was instead a case in which the US simply did not possess a feasible armament to employ. This claim is unsustainable, however, given the sheer evidence of the taboo identified above. In particular, the non-utility argument cannot explain why actors deliberately rejected offensive biowarfare options for reasons connected to the taboo. For example, non-utility cannot explain Leahy's vehement and highly moral rejection of biowarfare tactics against the proposition to target Japanese crops. Leahy did not oppose the plan on pragmatic grounds of practicality; his concern was explicitly a moralistic issue. Biological means of aggression were disparaged as an inherently taboo mode of violence that should not be utilized on the basis of that normative perception. Actors clearly and emotively articulated an opposition to these weapons that reflected taboo feeling and whereby even the *hypothetical* threat of biowarfare (regardless of weapon effectiveness) was dismissed as unacceptable.

The non-utility argument also cannot explain why the propositions of biowarfare that did occur were so heavily conditional and qualified. Biowarfare was strictly limited to defensive policies of retaliation. Notably, the sole Stevenson recommendation not adopted by the US government was the abandonment of a no-first-use policy. Furthermore, the types of biological attack that were suggested did not target humans. The Japanese crop proposal would not directly harm humans—and even this limited plan was still rejected as intolerable in respect to taboo feeling. The identified conditionality raises two key and related questions. First, why was the US not prepared to carry out smaller biological attacks involving the limited use of a biological agent, even if America was not capable of more extensive acts of biowarfare? Leahy's response indicates that this reticence was due to taboo belief. Second, why restrict conceptions of action to the defensive and non-human so as to better fall within the scope of taboo regardless of actual capacity if that normative expectation did not apply? Even the pragmatism of World War II could not fully undermine the entrenched sentiment that bioweapons comprised terrible devices to be used only under the most extreme circumstance. Moreover, even the most ominous extremes of World War II were never extreme enough to break the taboo in the form of even less extensive acts of biological aggression.

The bioweapons taboo also comprised a fundamental restraint on the decision-making process. Even actors who did not ascribe to the taboo as policy could not escape its normative expectation. The need to acknowledge the taboo regardless of one's own personal opinion only intensified as political and public actors were informed of the programme's existence (an issue that will be returned to in Chapter 6). The taboo was also influential through the construction of strategic narratives. Even those actors who sought an aggressive expansion of biowarfare options felt it necessary to refer and defer to the taboo with respect to their rhetorical expression. Strong political and public opinion in favour of the taboo again constrained US actors into either eschewing the topic entirely (e.g. Eisenhower's prohibition on all mention of biowarfare as the deliberate absence of narrative) or adopting modes of taboo talk (e.g. the PR campaigns endorsed by Forrestal and Stevenson). Ultimately, the use of bioweapons was never fully incorporated into national or military strategy even within a speculative or hypothetical sense. Policy and policy talk were limited to no-first-use and expressed the stringent rejection of offensive action in line with the taboo. World War II did constitute a point at which biological arms were in some ways standardized as a weapon for US military tactical employment in that a programme was approved. Yet these armaments were not integrated into aggressive political or military strategy within the time frame of this chapter. Indeed, US actors were even willing to walk away from the programme and the considerable $60 million investment in it once World War II was over.

Critically, the claims made within this chapter do not equate to a statement that the taboo was fully robust. Absolute normative adherence would be correlated with the absence of a US programme and this was clearly not the case here. The taboo was placed under significant strain as a consequence of military and bureaucratic dynamics. Specifically, the taboo was insufficient to entirely override the impact of two global conflicts. Yet the US programme should not be deemed a negation, or absence, of the taboo. Other pressures may have helped bring the bioweapons programme into existence and also later facilitated its revitalization in the context of the Cold War. Yet there also existed a distinct and credible recognition of the taboo. The UN declarations against biowarfare in particular clearly exemplify the presence of a taboo, whereby this sentiment was institutionalized in terms of the stigmatization, fetishization, and promotion of the taboo as policy. Such proscription may not have comprised the strongest version of taboo; the taboo, however, still exerted significant influence. And, as the fervent UN commitment to further prohibition demonstrates, this taboo belief was growing stronger in spite of international conflict. The Cold War was beginning to harden—but so was the taboo.

5
Allegation
Korean War

> This is one of the grossest falsehoods in history.
>
> **(Acheson 1952a: 424)**

A Beijing radio station made the first allegation on 22 March 1951: the US had used bioweapons as part of military action in the Korean War (Gross 1952a: 154). The report claimed that the Americans had deliberately employed smallpox (supposedly obtained from the Japanese[1]) against North Korea. The story spread rapidly throughout the Communist press.[2] North Korean Foreign Minister, Pak Hon-yong, followed up the claim with further accusations seven weeks later in an official cable to the UN Security Council (UNSC). He stated that US forces had utilized smallpox as a weapon as they withdrew from the Yalu River in December 1950. Pak branded the act a 'heinous crime' and demanded that America 'be brought to justice as a "bacteriological war criminal"' (CIA 1951a). When the new UN Commander, General Matthew Ridgway (in S. Harris 2002: 231), vehemently denounced the claims as 'deliberate lies', the Communist media derided his response. An article printed in the Hungarian newspaper *Izvestiya* rejected Ridgway's statement that the US was, in fact, attempting to rein in natural disease outbreaks within North Korea. The article mused 'whether it would not be nearer to the truth to say that he [Ridgway] received his decoration for spreading, instead of preventing, the epidemic' (CIA 1951b).

[1] Claims that the bioweapons came from Japan were possibly linked to the US' acquisition of Japanese-made vaccines as part of a public health campaign in South Korea, whereby Albert Cowdrey (1984: 160) comments that this 'was the germ of truth, so to speak, in the story'. Further Communist allegations of biowarfare were based on the 'evidence' that US tactics were similar to Japanese wartime experiments, such as the use of contaminated feathers. The implication of this claim was that US scientists had derived these tactics from their investigation of Japan's research, discussed in Chapter 4 (Cassell 1983: 270 fn54; Furmanksi and Wheelis 2009: 253).

[2] In April 1951, the Soviet Navy newspaper, *Red Fleet*, also alleged that the US had tested bioweapons on Canadian Eskimos in 1949, causing a plague epidemic ('Soviet Organ Sees Confusion in U.S.', 1951: 6).

The Biological Weapons Taboo. Michelle Bentley, Oxford University Press. © Michelle Bentley (2023).
DOI: 10.1093/oso/9780198892151.003.0006

The allegations resurfaced in full force the next year—at which point Ernest Gross (1952a: 154), head of the US delegation to the UN, said that 'the heavy guns of Soviet propaganda blasted out on germ warfare'. Pak introduced fresh claims at the UN on 22 February 1952. He alleged that American military planes had airdropped insects infected with plague and cholera over the region. Both Ridgway and Secretary of State, Dean Acheson, were again quick to deny. A US Department of State (DoS 1952a: 7) analysis reported that Acheson refuted the 'fantastic' claims of biowarfare. Acheson instead blamed any epidemic in North Korea on natural causes and the 'Communist inability to care for the health of the people under their control'. Premier of the People's Republic of China, Zhou Enlai, now publicly backed up the North Korean accusations and, in March 1952, extended them to China. Enlai inculpated the US for purportedly authorizing 448 aircraft to carry out 68 attacks throughout late February and early March aimed at Manchuria. These attacks again supposedly involved the dissemination of infectious insects as well as contaminated feathers for the purposes, said one Chinese newspaper editorial, of 'wrecking the armistice talks in Korea, prolonging and expanding the aggressive war in Korea, and instigating of new war' (in CIA 1952a: 1).

Other allegations claimed that the US had tested bioweapons on kidnapped North Korean and Chinese soldiers at Koje Island off South Korea (Parrott 1952a). The Soviet Academy of the Sciences also accused the American Geographical Society of producing a 'World Atlas of Disease' designed to help enact biowarfare and that this was already being used in relation to Korea. Comparing the map to Adolf Hitler's World War II medical experiments in the concentration camps, the Academy's journal claimed that the atlas was 'related to the reconnaissance and preparation of materials which can be used during the planned [biological] attacks on other countries' ('Reds Call', 1952: 5). The claims were again denied: UN Secretary-General, Trygve Lie (in Mayo 1953: 641), labelled the accusations as 'utterly false'; Ridgway (in Cortesi 1952: 5) swore by God as his witness that the claims were untrue; Acheson (1952b: 666) referred to 'the most false, continuous charges against the United States'; and Secretary of Defense, Robert Lovett (in 'Reds' Germ Charges', 1952: 2), viciously branded the statements to be 'absolute, complete black lies'. Truman (1952) also followed suit:

> The Kremlin cries that we have used germ warfare. There isn't a word of truth in that. We have never broken the Geneva convention in our operations in Korea. And they know that. They know it well. But they keep passing out the lies that have no foundation in fact whatever.
>
> **(Truman 1952)**

Deliberate lies

While 'for a few scholars the jury is still out' (Foot 1991: 425),[3] this study accepts that the Communist claims were false. Within the context of an analysis that seeks to ascertain the relevance of the bioweapons taboo, establishing the allegations' falsity is important in order to demonstrate that no major violation of the taboo took place, that is, no bioweapon deployment. First, it is unclear whether the US possessed the military capacity to carry out the specific types of biowarfare alleged. While analysts such as Endicott and Hagerman (1998: 186) argue that the US did possess the necessary bioweapons, their assessment is speculative and circumstantial. The US's own military assessment of the allegations dispute that the alleged attacks were possible (DoA 1977: 3–4; see also Hay 1999: 211; Leitenberg 2000: 166; Crane 2001). Others contended, however, that the US would have done a much better job of using bioweapons if they *had* chosen to engage in acts of biowarfare—carrying out even more destructive, efficient, and obvious attacks than those alleged (Clarke 1968: 24). Chisholm was quoted in a newspaper report:

> If the United Nations had used germ warfare, he [Chisholm] said, 'millions of people would die suddenly', it would be a far more 'effective' operation than painted in Communist propaganda and there would be *no mystery as to whether bacteriological weapons had or had not been employed* [emphasis added].
>
> **(Chisholm in 'U.N. Health Head', 1952: 3)**

Furthermore, America did not act like a party attempting to conceal any secret capability in relation to Korea. The Joint Chiefs of Staff (JCS) discussed political transparency on this point at a meeting held at the height of the allegations (DoS 1952b: 83). While the Director of Policy Planning for the State Department, Paul Nitze, dismissed Soviet demands for the US to be open and honest concerning its bioweapon capacity as 'silly', he also commented that: 'If it relates to their charges of bacteriological warfare, we might be better off to take the position that they can of course look at such things.' Nitze's statement contradicts any hypothetical argument that there existed some clandestine stockpile being utilized by the US against North Korea and China.

[3] E.g. Jaap van Ginneken (1977: 134); Jon Halliday (1984); Endicott and Hagerman (1998); Thomas Powell (2017, 2019).

Second, a related question mark hangs over the Communists' proof. US officials repeatedly rejected the accusations as containing serious 'flaws and inconsistencies' and based on 'phony evidence' (CIA 1952a: 1; DoS 1952c: 475). The US challenged the scientific basis of the claims; for example, in relation to the insects supposedly involved. The sheer number of different creatures alleged to have been used was suspicious. Gross (1952a: 157) commented to the UN on the astounding 'variety of germ-carriers [alleged to have been used] which would surely enrich any museum of natural history'. In addition, a 1952 CIA report stated that it was biologically impossible for some of the identified insects to carry the diseases in question. Communist news sources had claimed that the US used flies and fleas to transmit Bubonic plague. Of those two, however, only fleas can pass on the disease (CIA 1952b: 1).

The CIA also undermined reports concerning the insects' delivery. In one 'incident' it was claimed that insects were airdropped from 2,000 meters up over China and that 'within a matter of seconds a swarm of flies was found near the highway'. The CIA retorted: 'Now, 2,000 meters is approximately 6,500 feet—well over a mile. So assuming that "within a matter of seconds" was as long as 10 seconds, those flies were travelling well over *four hundred* miles an hour!' (CIA 1952b: 2–3). In a further negation, the CIA queried the meteorological conditions and season. Many of the attacks were said to have taken place during winter. Yet the insects would simply not have survived the cold:

> Perhaps the Communists want us to believe that these were super-insects able to withstand the sub-freezing temperatures of winter nights in Korea, or—as the *Manchester Guardian* expressed it, on March 8th—that these flies wore "special winter flight clothing—another triumph for U.S. logistics"!
>
> **(CIA 1952b: 4)**

US officials further argued that the insects and diseases cited were simply endemic to the area. Acheson repeatedly stated on this basis that the Communists had never been able to decisively establish that the presence of insects or any outbreak of sickness was unnatural ('Relief Handicaps', 1951: 3; Kaempffert 1952: E11). Overall then, the US maintained that the science did not add up. Yet Regis provides a further take on the evidence: it was *too* impressive. He comments that the sheer detail and quantity of proof was 'too good to be true' and 'lavish beyond belief': 'The real problem with the alleged evidence, however, was not that it was too weak but that it was too strong' (Regis 1999: 147, 227). The Communists had overplayed their hand.

There was a further controversy concerning certain 'confessions' extracted from American pilots captured during the conflict. The airmen had supposedly admitted that bioweapons were used. Washington officials strenuously rejected these confessions as the work of propaganda, brainwashing, and torture. Acheson (1952c: 777) publicly decried that it was 'crystal clear' that the admissions were 'trumped-up' falsehoods to the extent that their acquisition could be considered 'an international crime'. Charles Mayo (in 'Russia Accused by US', 1953: 1), US Representative to the UN General Assembly, officially presented the case for torture to the Political Committee and throughout the conflict would belabour this point: 'I need not belabor this point. The evidence of the falsity of such alleged confessions, and of the methods of duress and threat and wearing-down by which they are always obtained, is too well known' (Mayo 1953: 647). The US government also repeatedly pushed the line that no airman actually believed that their home nation was capable of biowarfare. This claim was backed up within the media. The *New York Times* reported one incident at a Chinese prisoner of war (PoW) camp in Weisong (MacGregor 1953). In trying to convince the PoWs that the US had carried out biowarfare, the guards brought in a glass holding a so-claimed 'infected' insect as evidence:

> Several curious prisoners looked the innocent bug in the eye, some of them somewhat suspicious that the Communists might have dusted it with a poisonous powder. But one G.I. calmly reached into the container, plucked out the insect and popped it into his mouth. The Communists in the room were at a complete loss.
>
> **(MacGregor 1953: 1–2)**

After two months in hospital the man was returned to his fellow prisoners:

> In the strictest confidence he related that he had been warned he would be killed if he disclosed he had not been the least bit ill at any time. ... That was the end of germ warfare propaganda at Weisong.
>
> **(MacGregor 1953: 1–2)**

Third, US officials cited the Communist opposition to an international investigation into the allegations as evidence that the claims were an untruth (Acheson 1952d: 529; CIA 1952a: 1). Acheson appealed to the International Red Cross (IRC) to examine the accusations in March 1952 (DoS 1952d: 452–3). The USSR rejected his call on the basis that the IRC was not neutral and was guilty of 'condoning the American actions' (Parrott 1952a: 1). Soviet representative to the UN, Jacob Malik (in 'Russians Demand', 1952: 9), further denounced the IRC as an 'organ covering up fascist atrocities'. Communist officials additionally alleged that the Americans only wanted a formal investigation in order to gain scientific data on their attacks' effectiveness, which the IRC was prepared to facilitate (Moreas 1952: 3). Paradoxically, the Soviets had previously made their own appeals to the IRC to intervene in respect of the biowarfare claims. Once an investigation became a realistic possibility, however, the Communists suddenly reversed their position. The US capitalized on this apparent hypocrisy to again condemn the biowarfare claims (Gross 1952a: 155). The Chinese elsewhere requested an investigation by the Soviet-affiliated World Peace Council (WPC). The International Scientific Commission (ISC)[4] considered eyewitness testimony, reports from (even Western) scientific experts, on-the-ground analysis, and the airman confessions. The report concluded that biowarfare had occurred. The ISC investigation, however, was dismissed by the US Psychological Strategy Board (PSB) as 'a complete fabrication' (PSB 1953: 1). America continued to protest its innocence; there had been no violation of the taboo.

The taboo and non-use during Korea

Accepting the allegations' falsity opens up the question: if America did not carry out biowarfare in Korea and China, why not? The explanation relates to the taboo. Biowarfare was understood as an act of extreme violation as well as political and moral regression. US officials recognized that any resort to bioviolence would be viewed as 'an international crime of the first magnitude' (van Courtland Moon 1999b: 72; see also Bunn 1969: 383). Gross (1952b: 33) passionately stated to the UNSC: 'The people of the United States, along with the rest of the decent world, are sickened at the very thought of the use of the weapons of mass destruction.' Furthermore, Roosevelt's no-first-use policy on bioweapons still held. The National Security Council (NSC 1950) had enshrined this stance in NSC-62 mere months before the Korean War commenced. The policy remained intact throughout the conflict and was reaffirmed in 1953 within NSC-147 (NSC 1953: 844). Milton Leitenberg (2008: 126) also refers to a letter from Truman to Congressperson Robert Kastenmeier (D-Wisconsin) on 25 July 1969, as evidence that the no-first-use policy was not violated at this time. The President stated in his correspondence: 'I wish to state categorically that I did not amend any Presidential

[4] The investigation's full name was: The International Scientific Commission for the Facts Concerning Bacterial Warfare in China and Korea.

order in force regarding biological weapons nor did I at any time give my approval to its use.'

America's non-use of bioweapons in Korea also reflected a critical *escalation* pertaining to the taboo at the international level. US actors were restrained by multilateral prohibitions opposing biowarfare that had been gaining traction and political weight at the UN since the 1945 WMD resolution. The concept of WMD was not merely a technical term. WMD comprised a highly emotive and norm-driven expression of the committed need to condemn and eliminate bioweapons. That concept was now incorporated across all aspects of arms control activity at the UN as an institutionalized construct. For example, the taboo underpinned the new UN Disarmament Commission (UNDC). The UNDC was established in 1952 under the UNSC remit. The organization was tasked with developing and overseeing the negotiation of a comprehensive arms control treaty. The conception of arms control employed here made explicit reference to WMD, where that term was then incorporated at the very centre of the UNDC's purpose (Bentley 2014: 50–3). Normative expectations against biowarfare were not only expanding in strength and relevance at this time, therefore, but also were accepted as a primary facet of the international community's bureaucratic structure via an intensive process of political stigmatization.

The way in which Korea was explicitly a UN action intensified the impact of the taboo's international normative institutionalization on US biowarfare policy. Korea constituted an inherently multilateral commitment based on UN partnership. Consequently, Korea simultaneously required and reinforced US efforts to be part of the UN, where bioweapons prohibition and the taboo now comprised a mainstay institutional project and an intrinsic value. The relevance of multilateralism here further connects to the US's commitment to the Geneva Protocol. As Representative Donald Fraser (D: Min; in USCong 1969a) said of the non-use of chemical weapons during the Korean conflict:

> On the use of gas in the Korean war, my understanding is that in part we were faced with the fact that we are alined [sic] with large numbers of other nations who are members of the United Nations, and who were also signatories to the [Geneva] protocol and had taken the view that any kind of gas would be a violation of the protocol, so that we were under constraints there . . .
>
> **(Fraser in USCong 1969a: 23)**

Fraser's view is extrapolated to biowarfare here on the grounds that this threat also relates to the Geneva Protocol. In combination with an on-going aversion towards biowarfare back home, therefore, there were also strong moral, normative, and attitudinal factors at play within the UN that related to Korea. Both domestically and as the product of international normative pressure, and regardless of the obvious pressures of international conflict, opinions concerning biowarfare continued to reflect the already-established belief that bioweapons were deeply pariah. This view again influenced American military strategy and tactical decision-making.

Big lies

US officials attributed the allegations to Communist propaganda.[5] Acheson (1952e: 428; see also CIA 1952a: 2; 'Germs and Propaganda', 1952: 14) repeatedly accused the Communists of deliberately manipulating the natural epidemics that were then raging through Korea. This tactic 'was too good a device to restrict to Korea', as one *New York Times* commentator put it in relation to the extension of the allegations to China (Waggoner 1952a: 1). Gross (1952a: 154) described the propaganda as a tactical response to political setbacks in Korea. Specifically, it was the case that the truce negotiations were going badly for the Communist side ('Eleven Weeks of Bad Faith', 1951: 30; Acheson 1952e: 427; Parrott 1952b: 1). A CIA (1952a: 2) report claimed that the Communists wanted to: discredit the US at the negotiating table with its allies and other non-Communist states; create fear that the US was deliberately aggravating the conflict, as well as East–West animosity more widely; undermine the US on a moral basis; strengthen the Communist moral position respectively; and deter further US military action. The accusations were portrayed as yet another example of the 'usual Communist "carrot-and-stick" approach of seeming concession combined with threats, demands and charges' (Baldwin 1952: 3). In particular, using the biowarfare accusations to publicly label the Americans as war criminals was considered the way to gain the upper hand in a situation of collapsing truce talks, not least where the US could then be presented as the villain (Parrott 1952b: 3). The Soviets had been accused of stalling the talks. Now it would be the US, and not the Communist side, who appeared to be standing in the way of negotiation and compromise.

[5] The allegations in respect of Korea were not the first of this type from Communist sources. Claims were made in 1950 that the US had carried out biowarfare in East Germany. The US was accused of air-dropping Colorado beetles ('Colorado Beetle', 1950: 10) and potato bugs ('Soviet Bug Charge', 1950: 3). The CIA dismissed these claims as a Communist fabrication designed to 'justify' the building up of an East German Air Force (CIA 1950: 3). UK representative to the UN, Sir Gladwyn Jebb, further said that these accusations were a cover-up for the failure of the potato crop in Czechoslovakia (DoS 1952e: 515).

The biowarfare issue also became embroiled in debates concerning PoWs. The Soviets had balked at a US proposal in 1952 that prisoners should be freed on equitable terms and on the basis of voluntary repatriations. Acheson (1970: 653) noted that the USSR's reticence was succeeded by new and highly 'lurid' accusations of further germ warfare. The Secretary said that the Soviets were making the claims explicitly to manipulate and muddy the political situation concerning PoWs with emotive claims of bioviolence. The claims were further employed as the basis of another form of political propaganda: public health campaigns within China, such as the Patriotic Hygiene Campaign. The campaign was instigated to restrict the spread of sickness throughout China at this time and the propaganda was seen as a means of mobilizing the Chinese population to comply with health measures (Chen 2009). The main slogan of the campaign boomed: 'Mobilise to promote hygiene, to reduce disease, to raise the level of the People's health, and to *smash the germ warfare of the American Imperialists!*' [emphasis added] (in Rogaski 2002: 387). In particular, the propaganda could be used to suggest that the outbreaks were not the consequence of any failing on China's part to secure the health of its people. The epidemics were portrayed not as the fault of the Chinese leadership but of enemy biological attack.

US officials also viewed the biowarfare claims as playing into a much wider propaganda that went beyond individual political spats or domestic political pressures.[6] The allegations were linked with a more expansive effort to undercut US credibility known colloquially within the US as 'hate America', or the 'big lie' (Kleitman 1950; Byfield 1951: 28).[7] The US argument claimed that the Soviets sought to undermine America through a comprehensive international campaign of calculated disinformation. The bioweapon allegations were presented as a particularly virulent aspect of this public relations exercise (CIA 1952a: 1, 1952c)—what the UN Commander after Ridgway, General Matthew Clark (1954: 203) termed the 'climax' of the propaganda.[8]

[6] Acheson provided 'evidence' of the unintentional publication of propaganda material prior to the first allegations to support his claims. Acheson (1952a: 424) said: 'This germ-war propaganda campaign was plainly planned well in advance. One of the Soviet publications jumped the gun and published a cartoon in which I was shown with a germ-war canister on my back. This was several months before the germ-war charge was launched. The mistake was quickly caught and the germ container disappeared in later editions. I suppose the editor did too.'

[7] US diplomats at the UN openly compared the campaign to Hitler's propaganda (Gross 1952a: 158; B. Cohen 1952a: 1030, 1952b: 295). Four key reasons were given for the 'big lie': to prepare the Communist public for new military action; disrupt security relationships between the US and Europe; frustration that the Communists had not made more progress in terms of their own power and to distract from this; and to address low domestic morale (Schwartz 1952: E7).

[8] General Clark (1954: 40) said of the allegations: 'This tactic, common to the Communists everywhere, is well illustrated by the old tale of the American engineers who visited Moscow. Their Russian hosts showed them the best the city had to offer in architectural and engineering accomplishments. But the very best was saved for last. That was the Metro (the Moscow subway). The Americans were astounded. They never had seen such a clean subway, so well appointed and so beautified with murals. Finally an American

US Ambassador to the USSR, George Kennan (of 'the long telegram' fame), sent a memorandum to the State Department on 22 May 1952, expressing his concern at the sheer extent and voracity of the biowarfare allegations. Recalling his vast experience with what he portrayed as the Soviet propaganda machine, he commented that he had 'never seen anything to equal in viciousness, shamelessness, mendacity and intensity what is now being done in this country to arouse hatred, revulsion and indignation with regard to Americans in general and our armed forces in particular' (Kennan 1952: 972). Accusations of biological aggression were, therefore, considered part of a highly public and rhetorical means to undercut the US on the global stage.

This claim of propaganda must be contextualized. The very concept of the big lie was, at least in part, a US construction. There were certainly actors who genuinely believed in the deliberate and systemic nature of Soviet propaganda—as evidenced by Kennan's memorandum. Yet there was also a clear incentive in demonizing the Communist position as a malicious fabrication. Calling out the bioweapons allegations as ideological warfare comprised a means by which to both dismiss any negative claims made against the US (by reducing the claims to false accusation) and to demonize the Communists for their resort to underhand tactics. The bigger the 'lie' being told, the simpler it would be to refute any accusation waged against the US. It is also the case that many Communists believed the biowarfare allegations. Even some Western actors would have seen truth in the claims. Crucially, the reference here to the need for contextualization does not negate the claim that the allegations were false. Furthermore, there still remains extensive evidence that the accusations were a political construction, that there existed an extensive effort to uphold them (e.g. creating false proof), and of deliberate and knowing collusion (Leitenberg 1998; Weathersby 1998). The evidence, however, is caveated here by the further claim that the US response to the allegations would still reflect the country's own political stance and advantage.

The bioweapons taboo as a strategic narrative

The bioweapons taboo comprised the key vehicle for the Communist claims. Reference to the taboo was intrinsic to the construction of the strategic narrative behind the allegations. In particular, the taboo was instrumental in communicating and highlighting the sheer abhorrence of biowarfare, as a

said, "This is wonderful, beautiful, much better than anything we have at home. But we have been here thirty minutes now and haven't seen a single train, or a customer. Tell me, when do the trains run?" "'Ah, ah, ah,' admonished the Russian host, "how about your lynchings in the South?'"

form of narrative strategy. Deliberately playing up the obnoxious character of biological armaments increased both the sense of emotive shock associated with the claims and, therefore, the vilification of the US for its supposed resort to such deplorable levels of behaviour. The worse the weapon, the worse the act of using it; and, consequently, the easier to demonize and politically undermine the US. In constructing the narrative, the taboo further helped to label the US in such stigmatized terms. The way in which the taboo (1) distinguishes biowarfare from conventional forms of aggression as a pariah form of violence, (2) demonstrates where the engagement with that abnormality represents a taboo act, and (3) utilizes the taboo to provide the language to label that repulsive deviation from the normal as an intolerable stigma, made the Communists' rhetorical attack compelling.

Communist actors utilized the taboo in order to link the US with perceptions of the disgusting. The narrative referred not only to the obvious disgust of disease but also to the contamination and intrinsic revulsion connected with insects as one of the purported modes of delivery. The same repulsion is evident in respect of the highly popular 'bug' movies of this time. Films include the 1954 *Them!*—where ants who have been exposed to radiation from nuclear weapons testing grow to enormous proportions, threaten to take over Los Angeles, and steal a lot of sugar. Admittedly, *Them!* demonstrates the ways in which these types of film tended to negatively link ideological issues around Communism with the threat of specifically *nuclear* arms. The fears of nuclear radiation, however, played into wider concerns surrounding contamination generally (Tsutsui 2007) and is connected here to biowarfare on that basis. Communist rhetoric effectively mirrored the disgusted fears that were exemplified within these movies. The employed narrative strategically sought to establish an association between America and repulsive creepy crawlies capable of spreading horrendous sickness.

The Communist strategic narrative also drew on issues of disgust-as-desire. As well as attracting global attention through the means of repulsion and stigmatization, the Communists' rhetorical appeal to the taboo exploited the fetishized interest of international policymakers. Who could not be intrigued by tales of covert biowarfare by a superpower? Not least claims that included such captivating stories as the use of weaponized insects, dangerous feathers, and even infected hamsters. The strategic rhetorical construction based on the taboo was explicitly fetishized and this narrative entrenched the issue on the international political agenda as a major talking point, if only at the status of gossip. The allegations created a sense of titillation which inspired extensive political interest. Such interest directly connects to biowarfare as a fetishized form of violence. It was explicitly the way that bioweapons are constructed as a fetishized concern that facilitated the attention paid to the Communist narrative.

Furthermore, the taboo's application as a strategic narrative enabled Communist actors to construct the biowarfare claims as an intrinsically moral issue. If biowarfare constituted an immoral act per the taboo, then America could be presented as similarly debauched by virtue of having engaged in that act. The Soviet media's language reflected this sentiment by emphasizing the unethical nature of the US's supposed actions. In June 1952, *Pravda* said of the allegations that the US was 'using the most inhuman barbaric means of warfare on a large scale' (in Gross 1952a: 156). Elsewhere the newspaper stated that America had utilized 'the criminal methods of mass homicide condemned by all honest men and banned by international conventions on poisonous substances, bacterial weapons, and napalm' (Gross 1952a: 157). *Izvestia* similarly described the situation as one wherein 'U.S. imperialists used the barbarous bacterial weapon which is condemned by the entire mankind and prohibited by the Geneva Protocol of 1925' (Gross 1952a: 156). *Trud* put it thus: 'The whole world learned with a shudder of new monstrous crimes of the overseas barbarians in violation of the most elementary principles of human moral behavior' (in Sprague 1952: 923). Even the airmen confessions were expressed within the moral language of the taboo and emotive references to WMD. First Lieutenant Floyd O'Neal (in WPC 1952: 569) 'said' of US biowarfare there was 'no need to use such terrible weapons of mass destruction against the civilian population of these countries'. He continued:

> Its [biowarfare] use threatens our very civilization. This type of warfare is against all known humanitarian principles. The peace loving peoples of every country condemn this type of warfare and the people who gave orders to use it.
>
> **(O'Neal in WPC 1952: 571)**

O'Neal's reference to civilians demonstrates the way in which the Communist narrative additionally drew upon the taboo construction of biowarfare as indiscriminate. A Soviet periodical stated: '[T]he imperialists do not shrink from any means of war against progressive mankind. ... The imperialists are reserving bacteriological warfare not only for the front lines, but mainly for the noncombatant population' (in CIA 1951c: 1). A further, more specific, accusation was made that non-combatant targeting was a deliberately racist act on the part of the US. America was accused of discriminately employing biowarfare against Asian populations in particular (Lubin 1952: 289; Waggoner 1952b: 3). Biowarfare was not simply an act of conflict but an illegitimate—and, according to *Pravda* (in Gross 1952a: 156), 'cannibalistic'—attack on the grounds of race. Once again, therefore, the Communist narrative portrayed biowarfare as an immoral act by an intrinsically unethical nation. The taboo facilitated this strategic characterization.

Immorality was also constructed as a question of arms control. Soviet officials attempted to exploit the UNDC as a political platform for their allegations (Waggoner 1952b: 3). In particular, the forum was used as a venue to publicly assert that the US had not lived up to the expectations of the Geneva Protocol. Despite the fact that America was not a signatory to the agreement, the Soviets still condemned the US for failing to respect the Protocol's terms. The claim comprised an accusation of immorality against the US in that the state was presented as unethical in failing to endorse the principles of established arms control measures pertaining to mass destructive weaponry. The narrative of immorality was not limited to Korea and China in this respect. Soviet claims went deeper: to accuse the US of possessing an intrinsic commitment to bioviolence that transgressed the moral expectations and values of the international community. Soviet actors argued that the US was deliberately attempting to shift the UNDC focus away from the issue of biowarfare in order to facilitate their own nefarious pursuit of these weapons, as well as opening up new opportunities to expand their nuclear capacity. The Soviets introduced a UNSC resolution on 26 June 1952, calling for all states to recognize and accede to the Protocol. US officials interpreted the move as yet another attempt to embarrass them internationally and to detract from American calls for an independent investigation into the allegations (DoS 1952e: 515).

Consequently, the ideals of the taboo were key to the 'big lie' as the strategic rhetorical expression of an intolerable form of behaviour. The taboo represented a basis on which the US was publicly identified as depraved and, therefore, beyond the realm of international acceptance. The taboo created a core and necessary moral standard that the US was deemed to have failed. To the extent that the Communist narrative presented the taboo as constituting one of the basic expectations of civilized international behaviour—indeed, of humanity itself—so the US could be politically condemned for failing to adhere to that expectation. The narrative approach, however, was not simply a painting of the US as an immoral actor. The strategy was also a means to disrupt Cold War alliances. A CIA (1952d) report stated at the time that the allegations were designed to cause 'the moral isolation of the United States from its allies'. The taboo was then central to both the strategic structure of the allegation rhetoric as well as the use of propaganda to shape the Cold War itself.

Retaliation in kind

US officials initially ignored the Communist allegations. American actors did not wish to inflame the situation or, as Deputy US Representative to the UN, Benjamin Cohen described it, to be 'pushed into a vilification contest' ('U.S. May Urge', 1952: 3). US embassy staff warned that any rebuttal would only draw unnecessary attention to and prolong the accusations (PSB 1953: 3). Officials also often assumed that the claims were too far-fetched for the international community to take seriously (Hamilton 1952a: E6).[9] One CIA (1952b: 1) report compared the allegations to the failure of Nazi propaganda 'due to wild overstatements carrying inaccuracies as obvious as those' promoted around biowarfare. Yet once it became clear that the Communist narrative showed no signs of 'dying down' (Hickerson 1952: 210), the US hit back with what the Soviets would call a 'slander campaign' of its own (CIA 1953). Wilson Compton (1952a), Director of the International Information Administration, said:

> Let us make no mistake—we are not going to get into a propaganda war. We are already in it and we had better stay in it, if we want what we need, namely, allies who understand us, respect us and have faith in us.
>
> **(Compton 1952a: 670)**

Kennan (1952: 976) said that there was 'a point' at which the US must respond to propaganda, however contrived that propaganda was. He believed that the biowarfare allegations constituted this point. Kennan 1952: 975) argued that the taboo 'abhorrence of mass destruction weapons' meant that a failure by the US both to reply to the allegations with a complete rejection of biological aggression, and also to confirm that the US would not employ or further develop these weapons, would make America 'vulnerable propagandistically'. Crucially, Kennan also recommended effectively embracing the taboo as a rhetorical means of reasserting American status in the face of the Soviet recrimination.

The taboo was desirable as the basis of a narrative strategy in that US officials did not want to reduce the allegations to a fight on the physical evidence. It was believed that it would be difficult and time-consuming to make their

[9] Secretary of State, John Foster Dulles (1955), recalled the situation in a presentation to the National War College: 'Going back to their bacteriological warfare charge against us in the Korean war—when that first came out I think a good many of us were inclined to think: Well, this is of no importance, and so forth, and so on, and laughed it off. It would have been much better if we had hit it right from the beginning.'

case by disputing the supposed facts of the case. While officials would eventually question the empirical validity of the Communists' evidence claims in public, the taboo initially provided a rhetorical focus by which to counter the Communist position that did not require talking about what actually happened in depth. Overall then, US actors now adopted the same rhetorical tactic as the Communists: they strategically utilized the taboo both to defend the American position and to disparage the Communists. The taboo was the enemy's trump card. It was a card that the US was now determined to play against them.

The first hand was dealt in respect of America's response to the Soviet's accusations at the UNDC. The US's discursive counterattack expressed a strong and highly public reaffirmation of its commitment to eliminate bioweapons and to the taboo. Dismissing the Communist claims as a 'monstrous falsehood',[10] Cohen (1952a: 1030) said: 'The assertion of the Soviet Union that the United States opposes the elimination of all weapons adaptable to mass destruction is untrue. The charge that we are against the elimination of atomic and germ warfare is not true.' Cohen accepted that the US *did* possess a bioweapon programme. He clarified, however, that this capability was for defensive purposes only. The programme was 'not for aggression, but to deter aggression' (Cohen 1952a: 1030). Cohen's focus on defence again relates to the qualified notions of the taboo as a non-offensive concept already discussed in Chapter 4. Highlighting that the US was not pursing biowarfare as an offensive tactic comprised a means of downplaying the programme and putting it back within the normative space of the taboo. Cohen (1953: 151) would later clarify that America recognized bioweapons disarmament as a principal aspect of the UNDC's remit. Again, the statement comprised a reassertion of the US's adherence to the taboo and the taboo's central role within international political thinking (see also Gross 1952b: 35).

Cohen additionally upheld the taboo as manifest within the Geneva Protocol. Cohen disputed Soviet condemnation of the US's non-ratification by proclaiming that America adhered to the Protocol's principles—in practice as well as in ethos. Cohen addressed the issue of why the US had not ratified the agreement. Cohen (1952e: 912) attributed American reluctance to a belief that the Protocol was insufficient to address the biowarfare problem and that 'paper pledges' of this type were unreliable: 'Bitter experience has taught us that such pledges by some states are valueless.' The comment was an unsubtle dig at the Soviets as 'some' state. Cohen was particularly keen to establish that any US opposition to the Protocol was not an opposition to the taboo:

[10] See also Cohen (1952c: 506, 1952d: 753).

'It would be a grave mistake to assume that, because the United States has not ratified the Geneva protocol, the United States is opposed to the general objective of the treaty, the effective outlawing of poisonous gas and biological weapons directed against human beings' (Cohen 1952b: 294). He argued that it was, in fact, the Communists who had violated the principles inherent to the Protocol. Cohen (1952e: 911) highlighted the reservations that the Soviets had endorsed in signing the Protocol. He then reasoned that the Soviets' conditionality demonstrated a condemnable lack of commitment to the taboo's ideals. In contrast, Cohen asserted that American ratification of the Protocol had not occurred because the US wanted a solution that was even *stronger* than the agreement. The US had not approved the Protocol because America desired a more substantial and more rigorous commitment to the taboo.

The US narrative further sought to undermine the Communists for simply *making* the accusation that America had employed bioweapons. The taboo made the biowarfare claims compelling. Yet the taboo also increased the controversy of those claims being made in the first place, especially if those allegations were then proven to constitute a fabrication. The taboo made the claim of biowarfare so serious a concern that, if false, the very act of accusation would be reprehensible itself. US officials repeatedly argued this point. Compton (1952b: 867) explicitly labelled the Communist allegations as an 'immoral propaganda' during a speech on the US-based international radio service *Voice of America*. Gross (1952b: 33) similarly branded the Soviets' supposed concern for the Protocol as a 'fraud'. As Acheson expressed in the epigraph for this chapter, the allegations were themselves described as gross. The intrinsic disgust of the taboo meant that any act which appeared to disrespect the norm could also be constructed as repulsive.

American officials presented the claims' falsity as a dangerous product of Communism itself. Mayo (1953: 647) denigrated Soviet political thinking on these grounds: 'I should like to suggest that the "germ warfare" propaganda campaign is the very type of activity that stems from the essential doctrines of Soviet Leninism and Stalinism concerning truth and morality.' Needless to say, Mayo was not complimentary on whether these doctrines were either truthful or moral.[11] The DoS (1952c: 476) similarly described the accusations as made 'by persons whose allegiance to the Communist cause outweighs their regard for truth and decency'. In expressing these claims, the

[11] Mayo (1953: 647) said that the Soviets 'have elaborated the Marxian doctrine to a point where no non-Communist can possibly perceive the "truth," and indeed truth is whatever the Communists decide it is. ... As to morality, Lenin put it very succinctly when he wrote: "We do not believe in eternal morality, and we expose all the fables about morality"'.

US strategic rhetoric was constructed as an act of othering. In contrast to the uncivilized nature of Communist behaviour and ideology, Mayo (1953: 647) argued that the US was defined by 'high moral principles'. Consequently, the taboo was again strategically presented as a type of moral indicator. US actors utilized the norm to establish America as superior in comparison to the taboo-breaking and, therefore, illegitimate Communist. The construction of illegitimacy was further employed as a means of discrediting the allegations. If the Communists could not be trusted, then their accusations could not be either.

Significantly, it was not only US actors who demonized the act of allegation. Sir Percy Spender (in Rosenthal 1953: 3), Australia's Ambassador to the US, described the charge of biowarfare as the 'most monstrous accusation that could be made by a state that calls itself civilized—whether it be so or not—against any other state'. Jules Moch (in Rosenthal 1952: 2), French representative to the UNDC, similarly commented that the allegations 'could only bring a "curl of disdain to one's lips"'. As such, the opposition to the Communist claims was not only a retaliation on the part of aggrieved American actors. There was an international recognition of the revulsion that related to the idea of accusation itself as unacceptable—albeit one that was expressed here within a fundamentally Western and non-Communist context by US allies. Wider opposition precluded the Soviets from winning points against the US on this matter in the UNSC, specifically in relation to the above Protocol resolution. The Soviet proposal was defeated 1–0 with 10 abstentions. Gross (in DoS 1952f) described the negative vote as:

> [A] measure of the scorn and of the repudiation which I think all ten members of the Council, except the Soviet representative, feel for the futile and vain trick which the Soviet Government has attempted to perpetrate upon this Council, in raising the false issue of the ratification of the Geneva Protocol.
>
> **(Gross in DoS 1952f: 38)**

Where other actors also refused to accept the legitimacy of the Soviet claims—and, further, where this illegitimacy was vilified explicitly because of its association with the taboo—so the USSR could not gain traction on this point, even where the resolution appeared to uphold ideals connected to the taboo.

There is one final strategic element to the US narrative: the rhetoric drew strongly on a theme of poison. Chapter 3 has already established that the concept of poison is a key aspect of the bioweapons taboo. High Commissioner, John McCloy (in DoS 1952g: 975), decried the Communists' tendency for 'poisonous propaganda'. Gross (1953: 476) portrayed the allegations as a case 'in which despotic governments use the weapons of secrecy, hatred, and fear to *poison* men's minds and to paralyze their will to resist tyranny' [emphasis added]. Cohen (1953: 151) accused the USSR of attempting to 'poison' both the UNDC and the US's reputation by presenting America as hostile to bioweapons' prohibition within that forum. The allegations were also described as venomous. General Ridgway (1952: 926) referred to 'the blind hatred, the *vituperative venom*, the vicious falsehoods, deliberately employed' [emphasis added] by the Communist side. Gross (1952a: 153) called the Soviet propaganda a 'venom which is being injected into the minds of men'. Consequently, US officials again rhetorically exploited some of the deepest and most repellent aspects of the taboo as the strategic means of attacking the Communist position.

Violation and propaganda

The biowarfare allegations also reinforced the taboo as a consequence of perceived norm breaking. This observation relates to the claim made in Chapter 2 that violation may not erode a taboo but can cement and intensify normative understanding. The increased awareness of, as well as a desire to protect and uphold, a norm that is associated with transgression can actually engrain a taboo within accepted thinking. According to this conceptual framework of understanding, the biowarfare allegations here both constituted a violation in the minds of those who genuinely believed the claims and introduced an idea of what it would mean to contravene taboo expectations even for those who did not accept the accusations' validity. The allegations set the situation up as one in which the notion of defying the taboo comprised a significant and immediate concern, if not proven fact. The perception of contravention effectively invited political actors to consider how they would respond to normative breakage and the implications associated with this, including at a reputational level. The international status of the claims exacerbated this feeling whereby bioviolence was given a prominent and fetishized place on the global political agenda as a taboo concern.

At one level, the accusations of violation acted as a reminder of the disgust associated with bioweapons. Continued uncertainty and non-use of bioweapons meant that the conceptualization of their effects was often still a product of imagination. The Communist claims, however, conferred the issue with a sense of reality, and subsequently of revulsion. The accusations outlined a new picture of what biowarfare could look like in real life. The reactivated sentiment of repulse associated with this perception reinforced both the taboo itself as well as international actors' acknowledgement of its normative expectations. The allegations and the US response additionally comprised a reminder of the behavioural connotations of this disgust—where these related to ideas of immorality and indiscriminate action connected to biowarfare as a repulsive act and the need to respect taboo principles.

At a further level, the sense of violation forced a wide-ranging and international recommitment to taboo. Political actors across the globe were required to respond to the supposed transgression. In doing so, these actors reasserted a belief in the taboo's validity. Cohen's speeches at the UNDC exemplify this trend as a public, official, and emphatic commitment to the taboo in direct rebuttal of Soviet accusations of contravention. Transgression demanded an unequivocal restatement of the international community's acceptance of, and adherence to, the taboo. The situation applied to: those making the allegations (where the act of accusation assumes compliance on the part of the accuser or risk criticisms of hypocrisy); those denying the claims (where they had to support the taboo in order to back up their protestations of compliance); and those on the sidelines within the international community (who would not want to be seen to condone transgressive behaviour in respect of biowarfare, including by others). Failing to express a clear endorsement of the taboo would have marked an actor out as deviant. The allegations, therefore, provoked an enthusiastic recommitment to the taboo which intensified its relevance as a normative construct.

The accusations also reinforced the taboo's stigmatization within UN bureaucratic systems. Soviet actions in bringing the claim of normative violation within the explicit remit of the UN (notably the UNDC and UNSC) highlighted the taboo as an institutionalized aspect of global politics. Bioweapons elimination as a taboo construct was now central to the ethos and institutional structure of the UN. Directly engaging that structure as part of the accusations both emphasized and further normalized the taboo within international political thinking. The process was self-reinforcing. Playing out these fights within the context of the UN's bureaucratic make-up buttressed the UN *as* the prime political site for such discussion and normative application. Actors' acknowledgement of this primacy then fed back into the international debate and development of the acceptance occurring at this time in respect of the taboo. By openly and forcefully engaging with the wider ideational and political processes towards WMD arms control, this activity reinforced both the taboo and its normative institutionalization.

The allegations and the US response to them were strategic. Consequently, it could be claimed that the taboo talk identified here constituted little more than cheap talk. The criticism would be that the Soviets only spoke of the taboo as they wished to damage the US. Similarly, the US only spoke of the taboo to the extent that they were forced to respond as a means of upholding their reputational position. It is reductionist, however, to assume that the references to taboo were entirely contrived from a perspective of normative commitment. While the taboo was manifest as a strategic narrative construct, its rhetorical employment still comprised the exploitation of a genuine norm—as seen, for example, in Fraser's identification of the Protocol as a restraint on the use of chemical and biological arms. It was not the case that the taboo was made-up or plucked out of thin air. The very reason why the taboo was utilized as a propaganda tool was because it had such weight with an international audience. If the taboo had not been normatively established to a significant degree, it would not have had value as a strategic narrative. The Communists drew upon this specific line of argument as they knew that this approach would resonate with other actors who also recognized the taboo as a legitimate concept. Indeed, these strategic narratives actually reinforced the taboo by upholding it as a moral standard for international political behaviour. The narratives solidified the taboo simply by virtue of the fact that it was so persistently and publicly articulated. Regardless of the strategic intention behind the rhetoric, therefore, these acts of normative restatement strengthened the taboo in ways that recognized it as a 'good' thing which demanded unqualified compliance and by which those who did not adhere were marked out as pariah. Korea gave airtime to the ideals at the heart of the taboo; indeed, the allegations placed those ideals at the very centre of the international community's attention.

Furthermore, US officials continued to recognize the taboo despite their concern that the Communists would capitalize on their allegations to pave the way for their own use of bioweapons (CIA 1951d, 1952a: 2; Barghoorn 1954: 45; Quarles 1954). The Communists were accused of preparing to carry out biowarfare in both a 'military and psychological' sense (DoS 1952h: 2; see also Gross in DoS 1952i: 1041; Gross 1952b: 36). A prominent article in *Science* magazine argued that Korea comprised the perfect opportunity for a Soviet 'trial' of such weaponry ('Germ Warfare in Korea?', 1950: 22). The Russians had recently tested their first nuclear device in 1949. As a logical follow-on, it was claimed that they would now seek to similarly exploit their bioweapons capacity (Clark 1954: 204). Moreover, the terms of the Geneva Protocol permitted bioweapon use in the form of reciprocal retaliation. If the Soviets could reasonably claim that the US had employed bioweapons, then this accusation could ostensibly validate a Soviet strike in kind (Hamilton 1952b: 3). The US was not party to the Protocol and so the rule of reciprocation was technically unnecessary. The justification of reciprocal use, however, would offer additional legal and political credence to the Soviet position—one that went beyond a simple eye for an eye. As Cohen (1952e: 912) said: 'With the Soviet practice of accusing others of what they themselves plan to do, we fear that they would have no qualms about making a false accusation and immediately thereafter launching a bacteriological warfare attack.'[12] Consequently, it was asserted that the allegations were a precursor to, and the political facilitation of, future Soviet biowarfare. Yet the US did not retaliate to that threat. Despite the apparent risk of Soviet biological attack, the US never seriously contemplated biowarfare, or even making the threat of bioweapons use. Continued American resistance to bioweapons indicates that more normative dynamics applied. The US still acted in accordance with the taboo.

First use of bioweapons (a caveat)

The US's continuing bioweapon programme should not be ignored. While Korea did highlight and enhance the taboo, the threat posed by the war would also spur on another domestic push in favour of an American biowarfare capacity within the military sector. Cold War politics were again cited to justify a need for the US to secure itself by all means available. As Robert Daniel Murphy (1954: 292), Assistant Secretary of State for International Organization Affairs, said in favour of the biowarfare programme at this time: 'Cold wars demand cold nerves.'

The Army Chief of Staff called for a full institutional review of biowarfare policy in May 1954. The review was established in particular to re-examine the no-first-use stance that was then still in place. The Chief of Staff in question was General Ridgway—who had done so much to deny the Communist allegations. Ridgway now demanded a review on the grounds that he believed a defensive-only biowarfare policy constituted 'an obstacle to the U.S. armed forces achieving a state of readiness against attack' (DoA 1977: 4-1-4-2). Furthermore, a review that ultimately supported first-use would assist in justifying a weapon programme whose very relevance was under threat at this time. Jeffrey Smart (1997: 48) portrays the situation as one in which 'two wars had come and gone without the employment of chemical and biological

[12] US military expert D.J. Dallen (1952: 30) also highlighted that anything the Soviets claimed negatively of others typically echoed what they wished to do themselves, or had done: 'It has become a kind of precept in Moscow to accuse the opposite side of deeds committed by its own agencies. … In so far as there is any truth in the Communist reports about infected "flies, fleas and rats" let loose in Korea and North China, these germ carriers undoubtedly stem from some peculiar "war industry" plants behind the Iron Curtain.'

weapons'. Why then bother continuing to invest in a biowarfare capacity if this arsenal was never realistically to be exploited? To quote one Air Force memo, bioweapons were considered by some to constitute 'an unwarranted luxury' (in Guillemin 2005: 101).

In addition to this questioning of the programme's strategic relevance, the CC was criticized for having made a number of undeliverable promises concerning bioweapons. For example, Creasy (1950: 8) had informed the Stevenson Committee that the CC could construct an emergency offensive bioweapon capability within only three months. The CC had then failed to develop any workable biowarfare response even years later. Creasy's apparent exaggeration may have comprised an attempt to promote the CC's value and organizational status (F. Smith 2014: 45). Yet his seemingly false claims would only encourage critics who could no longer see the point of maintaining a biowarfare programme. Ridgway's review was, therefore, a strike back by the biowarfare hard-line against their detractors. The CC and its supporters sought an endorsement of a first-use option that would enhance a perception of bioweapons' military utility, secure the programme behind them, and secure the CC itself (Goldman 2009: 534).

At the time of the review's establishment there was still considerable domestic resistance to a first-use policy. Yet feeling shifted in 1956 following a speech by Soviet Marshal and Minister of Defence, General Georgy Zhukov. Speaking to the 20th Communist Party of the Soviet Union Congress, Zhukov (in CIA 1957: 3) was deemed to have announced a new strategic commitment by the USSR to bioweapon use. Zhukov stated: 'Future war, if they [the West] unleash it, will be characterized by the massive use of air forces, various rocket weapons, and various means of mass destruction, such as atomic, thermonuclear, chemical and bacteriological weapons.' Zhukov was seen to have constructed a supposed Western biological threat in order to justify the Soviets' own use of biowarfare. The NSC adopted regulation NSC 5062/1 in response to Zhukov's statement. The policy abandoned a retaliation-only position:

> To the extent that the military effectiveness of the armed forces will be enhanced by their use, the United States will be prepared to use chemical and bacteriological weapons in general war. The decision as to their use will be made by the President.
> **(NSC 1956: 246)**

The new policy was officially incorporated into the 1956 edition of the US Army manual, *The Law of Land Warfare*. The amendment explicitly emphasized that the US was not party to any agreement restricting its capacity for first-use of bioweapons in order to pre-empt any criticism of the policy on those grounds (DoA 1956: para 38). Strategically, biowarfare was now technically on the table.

The policy shift was concurrent with new investment in the US biowarfare programme. The Department of Defense (DoD) responded to the threat of Korea by increasing the national biowarfare budget from $5.3m to $345m over three years, which Jeffrey Lockwood (2009: 163–4) describes as the 'phoenix of biological warfare' rising from the ashes. The programme underwent major advances as a result. For example, the 1950s and 1960s were a time of significantly expanded biowarfare testing, including open-air experiments on US cities (Bentley 2019a). This increase in testing, Newtol Press (1985: 95) wryly notes, 'made Fort Detrick the world's largest user of guinea pigs in 1960'. The CIA would also now engage in a series of highly confidential programmes investigating the use of biological agents to conduct mind control (Project Bluebird, Project Artichoke, and MKULTRA[13]—the latter of *Stranger Things* fame) as well as the stockpiling of such agents (MKNAOMI) (USCong 1976: 387–92). The CIA even actively considered bioviolence in respect of assassination plots. In one incident, officials seriously debated killing Egyptian President Gamal Abdel Nasser with botulism-laced cigarettes in 1956 (Shpiro 2009: 11).

A caveat to a caveat

The move towards an offensive policy ostensibly violates the bioweapons taboo. Yet there are three critical caveats to this statement. First, the US commitment to offensive action was highly limited. Biowarfare still existed as a fundamentally defensive concept. Even the CIA's controversial MK programmes and assassination plots were heavily restricted and founded within doctrines of defence. A 1976 US Senate investigation into MKULTRA and its associated programmes specifically noted that these projects began as an 'acute' defensive response to the USSR, China, and other Communist bloc countries (USCong 1976: 392). Bioweapon development in this case was primarily driven by the aim of uncovering what the enemy could do and what their agents were so as to ensure that the US possessed the requisite scientific and military capacity to defeat such an attack. Admittedly, the Senate investigation went on to accept that the Communist threat would later force the defensive response onto a more offensive footing (USCong 1976: 393). This

[13] MK is said to stand for 'Mind Kontrol' (Guyatt 2005: 42).

shift in policy would be accompanied by more aggressive research including illegal human experimentation that violated the participants' rights (USCong 1977). Defence, however, still remained an elemental concern even for those who favoured more belligerent measures. The defensive strategic preference is evidenced in a telling comment by General J.R. Rothschild (1964), former Commanding General of CC Research and Development, in his book *Tomorrow's Weapons*:

> Until I retired [in 1957] . . . I was not able to speak of chemical or biological weapons without prefacing my remarks with the statement that the enemy might use it. I was never able to speak of the offensive, only of the defensive.
>
> **(Rothschild 1964: xii)**

Even actors who desired an offensive option recognized that the mere *mention* of first-use aggression was taboo.[14] Offence was insupportable, irrespective of official shifts in policy. At the same time, the CC stepped up its propaganda of the previous chapter. The campaign would now even receive its own name: 'Operation Blue Skies' (Gould and Connell 1997: 103). CC officials again felt pressure to construct biowarfare policy within the frame of, and in response to, the taboo and its associated concepts such as defence-only weapons development. Consequently, it is of considerable doubt whether actual offensive use would have been politically or publicly sanctioned.

Furthermore, US military strategy was now increasingly focused on incapacitating bioweapons. Tactical arms were presented as significantly less offensive than strategic devices in that they would create, as the CC put it, 'war without death' (Hartke in 'Authorization of Appropriations', 1969: 23,220). Creasy gave testimony to a House Appropriations Committee in May 1958 on the supposed advantages of smaller forms of bioweapon ('U.S. Seeks Arms', 1958: 8).[15] He played up the attractiveness of biowarfare to an enemy on the basis that these weapons did not target physical infrastructure and so would appeal to a foe that wished to 'remove the human element: without destroying buildings and machinery'. Creasy concluded that America would be 'living

[14] Elinor Langer (1968: 121) provides a further example in relation to the CC and biowarfare: 'The Chemical Corps has a public relations problem even within the Pentagon. Defense Department employees, who have attended high-level Pentagon briefings on the whole range of U.S. weapons systems, report that seasoned generals and admirals who can calmly devote whole mornings (or whole careers) to the business of nuclear destruction quickly become queasy and indignant when it is the Chemical Corps' turn at the blackboard.'

[15] At a later committee hearing, Creasy was asked by the chair: 'Can you develop for the committee the appropriate roles in modern war of nuclear explosives, and chemical, biological and radiological weapons. … That is a big one.' Creasy replied: 'Yes; it is probably worth the price of a good lunch anyhow, but I will take a crack at it' (USCong 1959: 8).

in a fool's paradise' if it did not possess a like-for-like retaliatory capacity. He further stated that incapacitating weapons were the best means of developing such a deterrence within the context of limited wars. While Creasy phrased his argument in strategic terms, however, his reasoning was still constructed as a concession to the bioweapons taboo. Downplaying the mass destructive and immoral aspects of these weapons (however contrived that argument) was a move designed to make biowarfare more palatable to a wider audience still opposed to the perceived excesses of bioviolence. Policy was presented in terms that harked to the taboo: discriminate targeting, injury not death, and the relatively ethical means of wounding.

The second caveat is that Eisenhower continued to condemn biowarfare. He was asked at a 1960 press conference whether US bioweapons policy had shifted to a first-use stance. Eisenhower (1960a) replied: 'No such official suggestion has been made to me and so far as my own instinct is concerned is to not start such a thing as that first.' Eisenhower's reference to 'no official suggestion' has been cited to claim that the president was not aware of the policy change (Guillemin 2005: 108). The archival evidence disputes this argument. Eisenhower attended a meeting of the NSC a month prior to the authorization of NCS 5062/1. The minutes show that Eisenhower clearly knew that a first-use policy had been put forward. The proposal was addressed during the meeting and it was questioned whether this constituted a 'shift in policy' (Gleason 1956: 207–8). Eisenhower replied that, as he understood it, the policy change would only be introduced 'to encourage research and development in these weapons fields.' How informed Eisenhower was of NCS 5062/1 once it was officially enacted is another matter. Yet the President certainly understood that a policy was forthcoming.

Eisenhower's response at the NSC meeting demonstrates the critical point that he still viewed biowarfare and the new policy itself in purely defensive terms. Eisenhower would repeatedly reject first-use as an act not even to be contemplated. For example, in 1958 the Treasury asked for clarification on the US's bioweapons policy where there was concern 'that too little money was being spent on these weapons if we did plan to use them in war' (Gleason 1958: 249). The President:

> [O]bserved that what this government had always done with respect to these weapons was first of all to make sure that we had a good defense against their employment by the enemy and, second, that we had sufficient chemical and biological weapons to retaliate if the enemy used them on us. This was the policy, said the President, that he had lived with ever since 1918.
>
> **(Gleason 1958: 249)**

Consequently, Eisenhower consistently adhered to a retaliation-only position long after NCS 5062/1 came into force and on intrinsically confidential matters that were not subject to public scrutiny. Even behind closed doors, Eisenhower's position on biowarfare reflected the taboo.

The final caveat is that Eisenhower's rejection of a first-use option was not exclusive to the president. The very secrecy surrounding NCS 5062/1 indicates that military actors recognized that the policy would prove prohibitively unpopular if known more widely. Once again, the clandestine approach cannot be reduced to explanations of national security. Officials acknowledged that public and political audiences would not condone aggressive biowarfare, especially after the controversial and repellent allegations of Korea. Journalist Jack Raymond (1959: 3; 1960: 3) commented that the Pentagon 'recognized the popular aversion to "unorthodox" warfare' and that the post-Korea sentiment had 'handicapped' pro-biowarfare efforts. In a similar vein, Dr Richard Morse (in 'Germ War Defense', 1960: 3), Director of the Army Research and Development branch, rejected claims that America would ever employ bioweapons. In explaining this position, he said that 'in this particular area we must be more sensitive from a psychological viewpoint than you have any idea because of the wide variety of political, social and religious overtones'.

Critically, Morse did not approve of those overtones. Only two days later he was reported as decrying 'the psychological abhorrence of the term bacteriological warfare' and he recommended 'facing up to reality, getting the facts [that biowarfare constituted a viable means of military violence] to the world' (Morse in 'Army Conducts Nerve Gas Tests' 1960: 33). Morse's acceptance that this anti-biowarfare viewpoint existed, however, again demonstrates the presence and influence of the taboo. As a study on biowarfare by the Subcommittee on Disarmament of the Committee on Foreign Relations also asserted at this time: 'The moral aspects have an important bearing on the possible use of CBR [chemical, biological, and radiological] agents in future wars, however, and should be considered in relation to other factors' (USCong 1960: 24). Raymond (1965) would comment further five years later:

> 'CBR', the military shorthand for chemical, biological and radiological warfare, is emblematic of the most secret sectors of modern weapons research. And military men frankly explain that the secrecy is due not only to the potential havoc of CBR weapons but to the moral questions they raise. Nor has the development of more terrible weapons such as napalm and nuclear bombs seemed to lessen the revulsion.
>
> **(Raymond 1965: E4)**

Other political actors similarly took up the taboo cause. Congressperson Kastenmeier was an especially strong taboo advocate at this time. Kastenmeier expressed his deep concern that a shift to first-use had taken place in secret. He sought to extract a commitment from the US government and the military that this was not the case and that such a change would never be authorized—not knowing that his fears had already been already been realized (USCong 1969a: 136). Kastenmeier proposed a resolution to the House in September 1959 calling for an explicit no-use policy on biological and chemical arms (House Concurrent Resolution 433). Unsurprisingly, the resolution proved controversial in security circles. Kastenmeier received phone calls from CC representatives who attempted to convince him that bioweapons were in fact a beneficial and humane form of warfare (Hersh 1968a: 27). Kastenmeier offered to limit the resolution to a no-first-use statement in response to their pleas. The CC refused the proposed compromise as they wished to keep their options on the battlefield open. The Departments of State and Defense also opposed Kastenmeier's campaign. The latter attempted to reason that 'similar declarations might apply with equal pertinency across the entire weapons spectrum, and no reason is perceived why biological and chemical weapons should be singled out for this special declaration' (SIPRI 1973: 115). The resolution was assigned to the House Foreign Affairs Committee (HFAC), but no further action was taken.

Despite the resolution's failure, however, its proposal still expressed a significant reassertion of the taboo. Kastenmeier (1969: 10,271) made a statement to the House in which he explicitly referred to the first-use of bioweapons within the moral terms of the taboo. The Congressperson commented that the issue of an offensive biowarfare strategy was 'a problem which I feel is important to us as a nation of free people who honor and protect life, and hopefully, represent and live as a *moral and good people*' [emphasis added]. Kastenmeier also stressed the possibility that smaller nations would wish to develop biological arms in his speech. This argument would later come to comprise a call to arms for biowarfare advocates in that the risk ostensibly indicated that the US should develop a biological arsenal in order to retaliate or to deter. Yet here the argument was made with clear connections to the taboo. Kastenmeier said that the potential threat of bioweapons acquisition by smaller states demonstrated a need to uphold the taboo as a moral concern:

> This, of course, is why it is of paramount importance that the United States set a moral tone in this area. If we do not, we will have done much to initiate

> the production of these weapons by smaller nations who might otherwise have compunctions in manufacturing and using these weapons.
>
> (Kastenmeier (1969: 10,272)

Kastenmeier effectively argued that the taboo created a guardian against bioweapons use and a key means by which the biological threat could be lessened and controlled. Moreover, Kastenmeier (1969: 10,273) referenced the taboo as part of the US's moral identity in the face of Soviet propaganda, per Kennan's warnings. Kastenmeier contended that a first-use policy would engender a situation within which the US 'will have both added fuel to the propaganda mills of the Communists and strengthened their false charges about our use of such methods during wartime'. Kastenmeier continued in fetishized terms whereby he equated bioweapons with the very moral survival of the US:

> Mr Speaker, we are losing the propaganda battle, and unless we are willing to express publicly a moral national policy on this issue we will be creating the impression that we are oblivious to the horrors that these weapons bring if unleashed. I do not believe the American people want such a state to come into being. … We cannot allow the concept of survival to obliterate our spiritual and moral values, for if we do survival will have no meaning.
>
> (Kastenmeier 1969: 10,273)

Another step forward

The conflict between a first-use policy and the taboo of Kastenmeier's emotive appeal exemplifies the on-going tension between the pressures of conflict and normative expectation relating to biological aggression at this time. The decision to more fully incorporate offensive biowarfare into military strategy would ostensibly comprise a significant strike against the taboo. Especially in terms of stigmatization and codification, the shift in policy removed certain formal principles that had previously constituted a stronghold of the taboo in the form of a no-first-use commitment. NSC 5062/1 would seem to represent a substantial reversal away from the taboo beliefs that had underpinned the prohibition and rejection of bioweapons. Yet Korea also exacerbated deeply felt public and political commitments to the taboo. The taboo was a key factor concerning non-use by US actors in respect of the conflict. Normative feeling also formed the basis of a major international propaganda campaign on the part of the Communists, whereby those allegations subsequently reinforced the taboo within international political thinking. The taboo exerted the extensive power to scare and coerce—a quality which expresses an even more engrained aspect of normative acknowledgement than behavioural guidance alone. The taboo also provided the stigmatizing labels by which the US was rhetorically attacked, and by which the Soviet Union would be targeted in response. Furthermore, normative commitment was entrenched to an even greater extent by the allegations' expression through the UN.

It was also the case that bioweapons had never been more disgusting. Biowarfare was made visual and visceral in the form of highly fetishized tales as to what these weapons could achieve. Far from being referred to as 'just' another mode of violence, biological armaments were now palpably manifest within menacing stories of infectious insects. Whether those insects actually existed or not is irrelevant to the way in which they ascribed bioweapons a new sense of urgency and reality. Consequently, the Communists' lurid tales reactivated and reinforced the intrinsic repulse of the taboo. Korea also demonstrates the importance of disgust to the notion of a taboo and the way in which the norm distinguishes bioweapons from conventional forms of aggression. It was the capacity of bioweapons to make the skin crawl that underpinned the allegations as well as their political impact, which was significantly more extensive than any claim of conventional attack could have achieved.

Furthermore, the first-use policy must be contextualized. While the option of offensive action had technically now been incorporated within military strategy, the actual resort to first-use aggression was never realistically considered as viable by those capable of authorizing their use. Any opposition to the taboo was confined to a limited section of the military who could still not necessarily overcome the presidential sanction that first-use would require. Moreover, the pervasive secrecy surrounding the introduction of this military option says infinitely more about the prohibitive revulsion that this policy would inspire if knowledge of it ever got out than any need to secure national security. Why else did the military still need Operation Blue Skies? Because wider resistance to bioviolence was still substantial, per the taboo.

Overall then, the taboo was influential in relation to: the non-use of bioweapons; a highly limited and conditional incorporation of bioweapons into offensive strategy which never translated into actual employment; shaping both biowarfare policy and the extent of bioweapons development, including where these matters were still restricted by a preference for defensive measures; shaping political rhetoric, evident within both the strategic narratives of the USSR and the US with respect to biowarfare allegations, and also Operation Blue Skies; and indicating that both action and rhetoric were in large part characterized by a response to the international norm against biowarfare manifest at the UN and which received a considerable boost in terms of Korea. Critically, the claims made here are not intended to suggest that the taboo was now fully robust. Korea did not perfect the taboo. The normative force expressed here in respect of the taboo, however, remained sufficient to decisively influence decision-making, military action, and wider political behaviour within the international community. And while the domestic push towards offensive use was far from over within the American context at this point, bioweapons were now merely a matter of years away from being taken off the table completely—as the next chapter will demonstrate.

6
Renunciation
Nixon Administration

It started with a memo. Secretary of Defense, Melvin Laird (1969a), sent an official note to National Security Advisor, Henry Kissinger, on 30 April 1969 calling for a biowarfare policy review.[1] The memo stated that a review was required as 'the Administration is going to be under increasing fire as a result of numerous enquiries'. These 'enquires' referred to political campaigns within Congress opposing the US bioweapon programme by the likes of Congressperson McCarthy, a major proponent of the taboo. McCarthy (1970: 139–40) claimed the credit for persuading Laird to request a policy review. McCarthy said that he had met the Secretary at a White House reception and warned Laird of his plans to demand a major investigation into the biowarfare programme. Yet Laird was not sympathetic towards biological aggression either. Laird (in 'Authorization of Appropriations', 1969: 23,194) had previously supported a defensive biowarfare posture—a stance that he continued to express into the early days of the review process.[2] He was, however, long opposed to offensive employment and wanted the US programme reined in. Laird's background on the Health, Education, and Welfare Appropriations Subcommittee also underpinned a commitment to health promotion, which has been interpreted as a key aspect of his rejection of biowarfare (Frank 1975: 105).

Laird's was not the only request on Kissinger's desk. A review of the biowarfare programme had been attempted before. A study had been commissioned under President John F. Kennedy in 1963, but it 'evaporated' when Kennedy was assassinated only ten days after the review's authorization (Foster 1966a). A further effort was made under President Lyndon Johnson in 1966. This

[1] The review related to both chemical and biowarfare. Laird maintained, however, that these issues comprised two discrete concerns. Laird opposed use of the 'CBW term' due to the differences between the weapons and the distinct aims relating to the separate programmes associated with them ('Minutes of the Secretary of Defense's Staff Meeting', 1969: 2).

[2] For example, Laird initially said of the threat posed by Soviet bioweapons stockpiles: 'As much as we deplore this kind of weapon, if we want to make sure this weapon is never used we must have the capability to use it' (U.S. Chemical Arms a Deterrent', 1969: A4).

The Biological Weapons Taboo. Michelle Bentley, Oxford University Press. © Michelle Bentley (2023).
DOI: 10.1093/oso/9780198892151.003.0007

review was also disrupted, this time by President Richard Nixon's election (Goldman 2009: 538–9). NSC staffers saw the new Nixon administration as an opportunity to resurrect the process and issued another call for an investigation. The NSC subsequently reworked their demand to incorporate Laird's (Conway 1972: 49). Kissinger (1969a) sent Nixon a memo on 23 May proposing National Security Study Memoranda (NSSM) 59, which would set up 'a comprehensive review' of the biowarfare issue. Nixon assented and Kissinger (1969b: 1) instructed the NSC to proceed. The scope of the investigation was both to pass judgement on the future of the biowarfare programme and to recommend whether America should now ratify the Geneva Protocol.

Laird ensured that he would control the process from the start. Dale van Atta (2008: 295) said of Laird: 'His insistence that he be at the head of the parade rather than just on the reviewing stand made all the difference.' Laird was especially determined that the military would not dominate the proceedings. The JCS had already expressed its strong opposition to any curb on US biowarfare capability and Laird sought to spread the study thinly in order to effectively dilute the military's influence. NSSM-59 was then divided up across three intergovernmental subcommittees representing the military, diplomatic, and intelligence sectors. Additional input came from the President's Science Advisory Committee (PSAC). PSAC was an elite, independent, but somewhat secretive, body of 18 executive-appointed scientists. Each subcommittee would submit a report which would then be combined by the Interdepartmental Political-Military Group (IPMG). IPMG comprised members from State, Defense, the Arms Control Disarmament Agency (ACDA), and the intelligence community. The White House, however, would have the final say.

Kissinger's (1969a) memo to Nixon recommending NSSM-59 had already highlighted the 'substantial disagreement' over biowarfare between the disparate actors involved. Indeed, the repeated failure by these different factions to agree on a unified policy was already a key rationale for the review itself. This clash between the military and other agencies would define NSSM-59. For example, there was confusion at an Eighteen Nations disarmament conference held in Geneva in July 1969 over a statement issued by Nixon. Between the initial draft circulated to delegates and its official announcement a sentence was scandalously deleted. The sentence read: '[T]he specter of chemical and bacteriological warfare arouses horror and revulsion throughout the world' (Hamilton 1969: 8). McCarthy (1970: 143) recalled that he saw the original version with pleasure. When he contacted the White House press office to obtain a copy, however, he was sent one with the taboo-laden reference crossed out in black ink. A revised version of the statement—reissued for 'technical reasons'—was distributed three weeks later and included the contentious wording. This new statement served to underscore concerns that the mix-up was 'only the latest' consequence of problematic and on-going struggles within NSSM-59 (Hamilton 1969: 8). The initial omission may have been a simple administrative error in that both Nixon and a military representative had separately approved different drafts (McCarthy 1970: 144). Yet the very fact that there existed two versions to confuse demonstrates the bureaucratic conflict. Furthermore, McCarthy (1970: 45) speculated that the subsequent clarification took three weeks to be issued because there remained such uncertainty within Washington as to which position the US would adopt. The delay made it 'painfully clear that there was a major tug-of-war going on in the capital over the direction CBW [chemical and biological weapons] should take in the future'.

It was a war that the military would lose. The loss was in large part due to Laird's attempts to isolate and sideline the military's opinion. Over the summer of 1969 both the Office of Systems Analysis (OSA; a division within the Office of the Secretary of Defense) and PSAC submitted reports which opposed bioweapons on the grounds that these armaments lacked military utility. Laird concurred with their position[3]—and then proceeded to stack the deck by rejecting a parallel JCS assessment. The JCS report had been glowing in its estimation of bioweaponry and contested any suggestion of operational rollback. Indeed, the report recommended the programme's *expansion* regarding incapacitating and non-lethal agents. Laird simply removed the offending report from consideration. According to a memorandum by Deputy Assistant Secretary of Defense for International Affairs, Morton Halperin (1969: 1), Laird did so with 'the urging of Systems Analysis'.

Laird handed the responsibility for rewriting the report to the Office of the Assistant Secretary of Defense for International Security Affairs, headed up by G. Warren Nutter. Laird (1969b) stated in a memo to Nutter that he wished to see input 'in addition to the more purely military views'. The comment meant that Laird did not want Nutter's report to reflect the JCS pro-biowarfare stance. While this approach would not remove the military entirely from the NSSM-59 process, Laird effectively drowned out the key bureaucratic interests which supported a bioweapons capacity. Laird caused further controversy by extending the review's deadline in order to allow the

[3] Kissinger (1969c) observed in notes on a telephone discussion between himself and Laird on 18 November: 'Laird didn't think biological warfare is a strategic weapon.'

JCS position paper to be replaced. Even with the extension, however, there was little time to do so. Pressed to complete the report, the authors effectively plagiarized the document put forward by PSAC (Conway 1972: 141; Frank 1975: 119). At the same time, a secret memorandum from Laird to the NSC was leaked to the *New York Times* in October 1969. The Secretary had stated in the document that the US should 'stop producing biological agents for use in warfare' (in R. Smith 1969a: 1). The leak was assumed to have been carried out by Laird himself.

The IPMG outlined the following policy options in response to the submitted reports: (1) retain a full biowarfare capability including lethal agents (in relation to which the IPMG stated that the 'political costs would be unacceptable and there would be no offsetting military advantages'); (2) retain a limited capability for incapacitating agents (where the 'political costs would still be high'); (3) confine the programme to research and development for both offence and defence (the costs of which were again deemed 'high'); and (4) confine the programme to research and development for defence purposes only (no costs identified)[4] ('Analytical Summary', 1969: 3–4). Option 4 related to a fifth issue concerning a draft resolution submitted by the UK at the UN in what would ultimately become the BWC (discussed in Chapter 7). The report concluded that, if option 4 were selected, then 'the basic issue as to whether the U.S. can support the U.K. Draft Convention on BW [bioweapons] (with minor and understood alterations) is settled [i.e. the US would endorse the draft]'.

The die was already cast for option 4. On 17 November, the day prior to the NSC meeting at which NSSM-59 would be determined, Kissinger (1969d: 450–1) sent a note to Nixon outlining the IPMG recommendations. Kissinger strongly recommended that the President should support a policy of defensive research only. Kissinger (1969c) and Laird also conferred before the NSC meeting to ensure that the final decision went the right way. In particular, they discussed how best to neutralize the JCS Chair, General Earle Wheeler, who was expected to push for an offensive programme. Kissinger agreed to direct the President to call on Laird to speak first as a means of shaping the debate from the outset. Kissinger asserted decisively during the conversation that: 'There will be no affirmative decision for any BW program.' Consequently, the NSC meeting opened with a clear plot to isolate Wheeler and the pro-biowarfare position. Nixon (in 'Minutes of National Security Council Meeting', 1969: 360) commenced the discussion by referring negatively

[4] The report initially stated: 'OSD recommends this position, and all other agencies agree.' A handwritten note was added afterwards to clarify that 'all other agencies *except JCS* agree' [emphasis added].

to biowarfare as 'a difficult and unpleasant subject about which we have little real knowledge'. The statement could hardly inspire confidence in these armaments, or Nixon's confidence in them. Wheeler did still attempt to reason that bioweapons were relatively low cost and comprised a vital retaliatory capacity. His, however, was a lone voice.

Wheeler eventually capitulated and threw his weight behind option 4. Explaining why he was now prepared to sacrifice the programme, Wheeler said that the military did not 'feel as strongly about BW [bioweapons] as about CW [chemical weapons]' ('Minutes of National Security Council Meeting', 1969: 363). The military would fight unequivocally to retain a chemical warfare capacity; but they would not 'go to the wall' on bioweapons (Homan 1969: A1). Ultimately, the military knew that it was beaten. The Pentagon had already announced the previous week that it was abandoning plans to develop highly aggressive bioweapons specifically designed to have no antidote or vaccine (Raphael 1969a: 2). The decision was connected to a deeper recognition that they would not go to the wall on biowarfare because of the writing on it. Military officials believed that Laird was likely to win. The Pentagon's policy shift was an attempted compromise toward a more limited and less offensive programme. It was hoped that some capability could perhaps be salvaged if they agreed to reduce the programme's extent. The Pentagon's concession then indicates that the JCS foresaw that they were on the losing side. Laird later denied any animosity between himself and the JCS ('Laird Says Pentagon', 1969: 24). Yet Laird did still portray that same situation as one in which military interests had come into stark conflict with political ones: 'There are certain differences that exist when you look at anything from a military position without looking at the political considerations. They understand. So do I.'

Nixon (1969: 541) made an official statement on 25 November to unilaterally renounce bioweapons, to reaffirm a no-first-use policy on chemical weapons, and to limit biowarfare research to defence purposes only. The president further committed the US to both a resubmission of the Geneva Protocol for Congressional approval and to provisionally back the UK draft convention banning bioweapons.[5] Nixon decried the horrific and taboo nature of biowarfare in his speech. He said: 'Mankind already carries in its own hands too many of the seeds of its own destruction. By the examples we set today, we hope to contribute to an atmosphere of peace and understanding between nations and among men.'

[5] Nixon stated that US support was dependent on the need 'to clarify specific provisions of the draft to assure that necessary safeguards are included'.

NSSM-59, however, failed to clearly address the issue of toxins. The omission then raised a critical question: were toxins included in Nixon's rejection? Toxins are non-living organisms that are created biologically, for example snake venom. Consequently, toxins constitute an overlap between the biological and the chemical which has caused an ambiguity about which category they should fall into. Defence officials attempted to argue that toxins were chemical arms in the case of NSSM-59. Defining toxins as chemicals would then conveniently absolve them from the renunciation policy via this loophole. A confidential NSC memo to Kissinger sought to reassure the Advisor that the classification of toxins as chemical could easily be justified without contravening Nixon's stated position—although the same document further cautioned that toxins were more typically associated with the biological (Guhin 1969).

The NSC analysis was leaked to the press (DoD 1969). Kissinger (1969e: 2) spoke to Laird on the telephone about the memo's disclosure and the highly negative public response that it had received. Kissinger and Laird also noted the way in which toxins were now a 'big issue' within Congress as a consequence of the debate on the Protocol ratification. Laird commented that many people 'could care less [about the toxin issue] but it continues to be brought up'. Yet Laird was opposed to addressing the contentious concern with a 'big denouncement' on the following grounds: 'I don't think we should spill all of our guts before we get into this tear gas thing.' The reference to tear gas related to the controversy of US chemical agent use in Vietnam which was raging at the same time. Laird was keen to brush the matter of toxins under whatever carpet was available. Kissinger, however, disagreed that the administration could let the matter lie if pressure grew.

Pressure grew. On Valentine's Day 1970, Kissinger (in Naughton 1970: 1) publicly admitted that the failure to explicitly incorporate toxins within the renunciation policy was 'a slip up'. He had concluded that excluding toxins would undermine the political strategy of renunciation and risk the 'warmly' expressed goodwill generated by Nixon's announcement—as cautioned by Charles Yost (1970), US representative at the UN, in a telegram to the DoS. Yost's assessment additionally highlighted a further potential hazard pertaining to American credibility on this matter. Yost advised that 'if US were to retain bacteriological facilities for the manufacture of toxins, we would have difficulty in overcoming suspicion that we were in fact maintaining bacteriological weapons capability'. One unnamed official similarly added that all political support gained as a consequence of renunciation 'would go down the drain' if this situation occurred (R. Smith 1969b: 15). Consequently, toxins were formally included within the renunciation policy (White House

1970: 226). The news was poorly received by those working on the biowarfare programme and who were now facing unemployment. At Fort Detrick:

> There had been some gallows humor in the pilot plant at the time of official decommissioning among those afraid of losing their jobs. Someone had scrawled the words, 'Neck-stretching room' on the door of room 604. Inside the room, a rope with a noose at the end hung from an overhead pipe, and a sign on the wall said: 'The rope. If you must use the rope please leave a quarter on the tank. Purpose is to clean your blood up.'
>
> **(DoA 1977: 7–4)**

Further announcements followed concerning the programme's dismantlement. In January 1971, Nixon (1971a) revealed that Pine Bluff Arsenal would be given over to the Food and Drug Administration; and on 18 October, that Fort Detrick was to become a cancer research facility (Nixon 1971b).

The taboo and military utility

Renunciation has frequently been explained in terms of military utility (Rose 1972: 10; Christopher et al. 2001: 29–30; Tucker 2002; Beard 2007: 276; Fidler and Gostin 2007: 47). The argument claims that bioweapons were considered too extreme, difficult to deliver, and uncontrollable in respect of their destructive effect to be militarily effective. The armaments were unusable and, therefore, dispensable. US Ambassador to the UN, James Leonard (in Mangold and Goldberg 1999: 55), commented: 'I can't stress this enough. The US military did not think there was any military use for biological weapons. This was the main reason for allowing the BW [bioweapons] ban to go ahead.' The utility argument ostensibly informed the OSA and PSAC position papers which played such a key role in the NSSM-59 process. Paul Conway (1972: 139–40) argues that the studies directly influenced Laird and inspired his later belief that 'biological weapons could be sacrificed'. Conway states in particular that the documents explain why Laird had spoken out in favour of a deterrence capacity even immediately prior to reading the reports and then abandoned this position in preference to full elimination afterwards. Why retain a weapon that you would never employ? The high likelihood that US actors would never resort to biowarfare tactics meant that there was no point even possessing these weapons as a deterrence option. Nixon (in Safire 1998: 22) reportedly said: 'We'll never use the damn germs so what good is biological warfare as a deterrent?' Nixon further stated that bioweapons were unnecessary as long as the US possessed nuclear arms: 'If somebody uses germs on us we'll nuke 'em' (Safire 1998: 22).

Actors further argued that the US could exploit the utility argument in order to prevent others from developing a biowarfare capacity, especially smaller and weaker nations (Roffey, Tegnell, and Elgh 2002: 451–2). If the US could establish that bioweapons were ineffective, then the rest of the world could be convinced to disarm for the same reason. An HFAC report stated that successfully demonstrating to others that biowarfare could not be considered a viable option would remove a potential threat to the US. That threat could otherwise see biological agents 'fall into the hands of a future Hitler' (USCong 1970: 7).

Three related problems exist with the utility argument. First, it can be counter-claimed that renunciation put America at a military disadvantage. Eliminating any weapon irrespective of its utility could be viewed as an act of insecurity; for example, where others may then develop that same weapon in order to secure an advantage. Indeed, the USSR reacted to US renunciation by expanding its bioweapon programme in what would become the infamous *Biopreparat* (Alibek 1999; Leitenberg, Zilinskas, and Kuhn 2012; Zilinskas 2016). Steven Allen (2007: 10) opines on this point that renunciation was 'either visionary or suicidal'. Renunciation can also be characterized as an excessive response to the issue of utility. Matthew Meselson was a government scientific advisor and a key player in the anti-biowarfare faction during NSSM-59. Meselson (1999: 53) said on the matter of renunciation: 'It is important to note that these US decisions went far beyond the mere cancellation of a program. They renounced, without prior conditions, even the option to have biological and chemical weapons.' Similarly, a SIPRI (1971c: 281) report highlights that the decision to renounce was considered 'as the only true disarmament measure that had been taken during the whole post-war period, and as the only decision involving any measure of military sacrifice on the part of superpowers'. While such an extensive commitment may be explainable and justifiable from the perspective of a taboo, the US's full strategic rejection of bioweapons could be argued to constitute a disproportionate response to a simple calculation of military viability.

Similar question marks hang over other justifications for renunciation which pertain to utility. If renunciation was designed to demonstrate bioweapons' ineffectiveness to others, why would US officials then bother to advocate for an international ban in the form of the BWC? Why would an official prohibition be needed if everyone now accepted the lack of utility as reason enough for abstention? It is also unclear that the argument of a nuclear substitute to bioweaponry holds water. Kissinger (1969f: 4) elsewhere defended the retention of chemical weapons as he feared that chemical disarmament would encourage a dangerous reliance on nuclear arms. Kissinger's argument expresses the exact opposite logic to the one supposedly applied to bioweapons. Indeed, Laird ('Laird Defends', 1969: 4) openly stated he would not wish to depend on the nuclear option as an alternative to bioweapons. America had possessed a nuclear capacity since 1945 and this arsenal had never served to halt or significantly qualify the bioweapon programme (even if bioweapons had frequently been sidelined as something of a less alluring second cousin to nuclear arms). Moreover, the JCS and CC had now promoted a focus on smaller and incapacitating bioweapons. It is therefore difficult to see that strategic nuclear weapons would provide an appropriate and proportional replacement for these much lesser forms of destructive agent. In addition to such counterarguments, the US had now signed the 1968 Nuclear Non-Proliferation Treaty (NPT). Yet nuclear elimination would demand that the US abandon the very weapon that was supposed to substitute for a biowarfare capacity. If renunciation were premised on an increased reliance on nuclear weapons, it cannot also be the case that the US was actively prepared to restrain its nuclear arsenals. It is also the case that if bioweapons were militarily useless why would you even require a proxy (nuclear or otherwise) for that which is not worth anything? Why substitute nothing?

Second, there *was* evidence of bioweapons' military viability. Despite Kennedy's (1960: 64) assertion that biological armaments were 'dread weapons', he (and later Johnson) had overseen significant increases in biowarfare funding and research. The biowarfare programme was expanded following a military review by Secretary of Defense, Robert McNamara, under Project 112 (DoA 1977: 5-3, 5-5). The programme's enlarged investment meant that bioweapon development had moved forward considerably. Figures at Fort Detrick found the claims that bioweapons were too unpredictable and unreliable 'especially galling' and an insult to their research (Regis 1999: 208). US scientists believed that they had made significant advances in the biowarfare field, which could justify a continued programme.

Third, many academic studies on renunciation recognize that utility is not a sufficient explanatory factor. Nixon's decision is elsewhere presented as an intrinsically political decision that related to one or more of the following additional issues: chemical agent use in Vietnam and similar scandals (Goldman 2009: 537–48; Revill 2018: 392–3); domestic pressure in the form of Congressional actors and scientific advocates (Conway 1972: 183); international coercion (Revill 2018: 394); Nixon's desire for a positive pubic image as a 'man of peace' (Tucker 2002 128); US–USSR nuclear diplomacy (Mangold and Goldberg 1999: 55; Goldman 2009: 568; Revill 2018: 392); as well as policies of détente (Morris 1977: 99). Conway (1972: 183) further adds that renunciation required 'a dash of luck'. Yet this wider analytic discussion still overlooks a vital dynamic: the bioweapons taboo.[6] The available studies have often negated the taboo as a causal factor by rejecting the relevance of moral impulse. For example, Jonathan Tucker (2002: 127) states: 'Nixon did not view biological warfare as a moral issue but rather as a military and political one.' Al Mauroni (2000: 51) similarly dismisses any claim that Nixon was compelled by moral restraint as a media fantasy. He attributes the President's decision instead to 'partisan political hotdogging' designed to curb Democrat opposition. Ignoring the taboo, however, is deeply problematic. First, it is unclear that moral reasoning did not comprise an explicative factor. For example, Nixon justified renunciation 'as a reflection of his own, Quaker attitude toward weapons of mass destruction' (Binder 1978: 18). Second, (apart from luck) *all of the above factors can be reinterpreted as connected to the taboo.*

Even the utility argument can be re-examined in respect of the taboo. Bioweapons were rejected in utility terms on the basis of indiscriminate unpredictability and unreliability of effect. The same constructions of destructive impact, however, are also central to the taboo, per the discussion of Mueller in Chapter 2. Opposition to the extreme killing of non-combatants within an uncontrolled strike reflects the moral repulsion of the taboo against biowarfare. Concepts of utility pertaining to this issue were publicly expressed as taboo talk. For example, Secretary of State, William Rogers (in USCong 1971: 40), said during Congressional hearings on the Protocol that renunciation had occurred after it became clear that the US would never employ bioweapons even in like-for-like retaliation. Rogers stated that this was the case 'because biological weapons have such a terrible permeating impact on the civilian population'. While portrayed as a practical assessment of deterrence, therefore, Roger's statement also reveals the extent to which the opposition to indiscriminate force (a taboo concept) guided decision-making. Rogers further drew upon the taboo in labelling bioweapons 'a horror'.

Key taboo advocates also deliberately contrived utility arguments as part of a strategic narrative to promote the anti-biowarfare stance. Advocates cited utility claims—but only as a tactic to uphold the taboo where this strategic approach was considered more persuasive than direct appeals to moral reason. How else would you talk to the likes of Wheeler, who was never likely

[6] James Revill (2018: 388–9) mentions the taboo in his study but does not apply the concept directly to the policy or connect this to Nixon's decision-making.

to be swayed by claims of ethical sentiment? Military supporters would be best convinced by rhetorical means that reflected their own terms of thinking. They needed to be beaten at their own game. Framing the elimination of bioweapons as a logical military calculation provided the basis on which to challenge pro-biowarfare figures even if this argument was irrelevant or false to the constructing actor. Chair of the PSAC report, Ivan Bennett, was a staunch taboo advocate. Yet Bennett employed a utility line of argument in order to make his case. Speaking at an NAS symposium on chemical and biowarfare chaired by Meselson in October 1969, Bennett (1970: 271) explained that the best way to gain approval for the biowarfare programme's elimination was by undermining the utility claims on a practical basis as opposed to making moral arguments:

> Arguments based upon morality alone have little impact upon the Pentagon or ministries of defense anywhere. To a large extent, the political decisions that will be required to further restrict these [biological] or other forms of weaponry or to abolish them will be dependent upon convincing military establishments that they can relinquish an 'option' by arguing within the framework of their responsibilities and *missions as they view them*. The task of translating convincingly the moral argument of the threat to civilians implied by non-controllability into a military argument of technical unreliability of a weapon system is an example, and a simple one at that, of an analytic approach to problems of disarmament [emphasis in original].
>
> **(Bennett 1970: 271)**[7]

Consequently, the claims that the PSAC report adhered to the utility argument are misplaced. The document can in fact be interpreted as a manipulated expression of the taboo. Bennett did not reject biowarfare on the grounds of utility. He merely believed that this specific line of argument comprised the most apposite means of achieving a rejection of biowarfare that was actually driven by taboo sentiment. The tactic was not exclusive to Bennett. Meselson (in Rosenbaum 1969: 17) elsewhere outlined his adoption of the same strategy. Meselson was similarly cognizant that claims to the taboo would be controversial and uncompelling to a military audience. In response, he stated that he had instead focused on the undesirability of indiscriminate force as this would comprise 'the approach of a military planner'. The utility argument was then merely a cover for, and a strategic facilitation of, the taboo. The contrived narrative constituted a method of making the taboo

[7] Bennett (in USCong 1969a: 60–1) repeated this statement with minor changes at Congressional hearings on biowarfare policy a week prior to Nixon's renunciation announcement.

position more palatable to a hard-line audience. It was, however, still the taboo that comprised the key motivational factor.

Vietnam and other animals

The taboo also relates to the controversy concerning the US's employment of chemical defoliant and riot agent in Vietnam. Excessive conventional violence enacted during the Vietnam conflict also had inspired uproar and denunciation; for example, the 1968 My Lai massacre in which American troops indiscriminately killed over 400 Vietnamese civilians (Oliver 2003). Using chemical agent, however, was seen to constitute an especially contentious and reprehensible act. The claims of chemical employment were exemplified by the dissemination of Agent Orange. Agent Orange is an infamous anti-plant agent that destroyed not only flora, but also food sources and Vietnamese lives—inflicting localized famine, long-term health conditions, birth defects, and death. America attracted widespread condemnation across public and political audiences for these actions (Bentley 2019b).

The controversy was extended to biowarfare due to concerns that America would expand their chemical tactics to the biological ('Research Memorandum' 1962: 476; CIA 1969a). Critics argued that if the US was prepared to employ chemical agent in Vietnam, then the country was similarly capable of biological aggression. There was also concern that the Communists would exploit the same connection between chemical and biological agents. Recalling the biowarfare allegations pertaining to Korea (discussed in Chapter 5), US officials raised a red flag that Communist factions would choose to capitalize on American chemical agent use in Vietnam in order to then disparage the US's biowarfare capacity (Neilson 1961: 642). Secretary of State Rusk (1962: 607), cautioned that Communist accusations regarding US chemical employment in Vietnam 'could give rise to increased charges of use of poison gas and perhaps even of biological weapons against Asian population [sic]'.

The hostile response to America's actual and potential tactics in Southeast Asia was more than simple opposition to a violent act within a contentious conflict. The concern expressed comprised a deeply moral revulsion towards biological aggression engrained within a taboo-driven rejection of usage and the belief that this type of belligerence would uniquely exceed the rules of war. US officials acknowledged that their actions were subject to highly potent social standards opposing biological warfare. Edward Murrow (1962:

240), Director of the US Information Agency (USIA), commented: 'Chemical and biological warfare are subjects which arouse emotional reactions at least as intense as those aroused by nuclear warfare.' A *New York Times* article highlighted a further aspect of the taboo in respect to the conflict ('Gas (Nonlethal)', 1965: 42). The newspaper described the way in which Vietnam had been portrayed as a moral war whereby the US claimed that it was fighting to contain the evils of Communism. Echoing Just War theory, the article contended that the perceived morality of the conflict created an expectation that the US should act in ways that were also principled. Biological violence would be seen to violate this moral standard. If the US were the 'good' guys in Vietnam, then they could not also legitimately contravene the taboo by sanctioning the use of a 'bad' weapon.

The taboo was further evident in respect of the Geneva Protocol. US officials never put forward the obvious argument that their country had not ratified the Protocol and so was not technically obliged to adhere to its principles (Bunn 1970: 198). The US could not violate the Protocol regardless of whether it had employed chemical or biological weapons in Vietnam. Officials could potentially, therefore, have justified their position in Vietnam on the grounds of non-ratification. Yet the US response never challenged the Protocol's relevance or the ideals of the taboo inherent in it by doing so. The US maintained that it adhered to the Protocol irrespective of the refusal to ratify. As the IPMG report recognized, 'the principles of the Protocol have become part of customary international law' (Analytical Summary', 1969: 45). Indeed, there is no evidence that America ever contravened the terms of the Protocol by using bioweapons in the Vietnam conflict. US actors did accept that chemical agents had been employed. Officials clarified this position, however, by stating that this action did not qualify as a transgression of the Protocol (e.g. USCong 1970: 21). It was asserted that the agents used did not fall within the restriction of the agreement by virtue of their non-lethal status. America had not broken the rules.

The focus on weapons that could be deemed to fall below a normative threshold of the taboo also comprised a feature of US rhetoric elsewhere. Specifically, military actors once again referred to taboo-sensitive claims that nonlethal bioweapons were more moral and acceptable (see Chapter 5). Incapacitating agents continued to be sold as an intrinsically humane tactic, whereby these armaments were presented as less offensive than even conventional means of violence. As such, the weapons were also constructed as less taboo than their strategic counterparts. 'INCAP'—short for 'incapacitating'—became a zeitgeist expression of the supposedly benign nature of these armaments (R. Smith 1969c: 4). Military officials similarly attempted to popularize the phrase 'Benevolent Incapacitators' in a further bid to make these agents sound less repugnant (Reston 1965: 42). Once again, the taboo drove the terms of debate and rhetorical presentation.

Yet US officials also knew that they were losing the pro-INCAP argument. Descriptions of the more civilized impact of non-lethal weapons were designed to appeal to the taboo position. The taboo, however, was too strong to maintain the claim. Secretary Rogers (1969) eventually sent a memorandum to the US Mission at the UN advising against any further reference to benevolent usage. Rogers felt that the US could no longer adequately defend itself on those grounds given the destructive evidence of on-going chemical agent use in Vietnam. The idea that biowarfare could be classified as humane was also challenged at the domestic level. Representative Donald Fraser (D: Minnesota; in USCong 1969a: 3) commented during a Congressional hearing: 'The ultimate myth employed by CBW [chemical and biological weapons] advocates is to justify their development and use as "humane" weapons.' Rosebury also again spoke up. He publicly responded to Rothschild (1968: 32), who had written a letter to the *New York Times*. Rothschild had questioned the editorial integrity of the newspaper for referring to biowarfare using, what Rothschild claimed were, 'unreasoning … emotion-triggering words' such as 'terrifying'. Rothschild argued that this expression constituted an inappropriate means of describing humane weapons. Rosebury (1968: E19) explicitly dismissed Rothschild's references to the humane: 'At best it suggests a specious relativity: we are told, having been punched in the nose, to be grateful that we haven't had our teeth knocked out. Only the puncher thinks in such terms.' Rosebury also prioritized the taboo above all other military considerations: 'I suggest that we may need to accept the fact of popular revulsion to certain weapons, however unreasonable it may seem to us [i.e. to the military position].' The scientist asserted that the taboo was too compelling to ignore.

Scandal on top of scandal

Vietnam was not the only scandal. The 'other animals' were 6,000 sheep that were killed in March 1968 at Skull Valley in Utah. The sheep died after VX nerve agent was accidentally released from a nearby biowarfare testing facility, Dugway Proving Ground. The Army initially denied any wrongdoing. Congressional investigations later extracted—or rather 'wrung … a shred at a time' (Reed 1969: 14)—a confession. Military officials admitted that the spraying device on a testing plane had malfunctioned mid-flight and caused the agent to disperse. The next day the sheep were 'acting in a most peculiar fashion—dazed, staggering, jerking their necks spasmodically to the side, and finally dropping to the ground, apparently exhausted and unable to rise' (USCong 1969b: 210). The Congressional hearings were held in May 1969. The date coincided with Kissinger and Nixon's commissioning of NSSM-59. Dugway, therefore, put significant pressure on political actors, their decision to carry out a policy review, as well as the review's outcome.

The Congressional hearings also oversaw a public commitment to the no-first-use of bioweapons by the Pentagon. Chair of the investigating committee, Henry Reuss (D: Wisconsin; in USCong 1969b: 126), questioned K.C. Emerson, Acting Deputy Assistant of the Army (Research and Development). During the testimony Reuss asked: 'As I understand it we [the US] pledged ourselves not to use it [biowarfare] first and somebody else will have to use it first; is that correct?' Emerson replied: 'That's correct.'[8] The comment directly contradicted NSC 5062/1. Emerson did not explain the rationale behind his critical statement. The clear rejection of a first-use option indicates, however, that officials did not believe in, did not know of, or were not prepared to publicly acknowledge the US's policy of offensive action. Any one of these three options would be sufficient to demonstrate that there existed a prohibitory reticence towards the offensive use of bioweapons. Even if Emerson's position was only the concealment of an active policy, his actions would still reveal the desire to keep secret what was publicly intolerable whereby offensive biowarfare was recognized to be a taboo. Consequently, the hearings evidence both a direct refutation of the first-strike policy and the impact that public and political outrage had in pushing forward elimination efforts. Each of these two factors reflect the expectations of the taboo.

Scandal multiplied. The Army was caught disposing of unwanted chemical agent by loading the agent onto ships and then scuttling the vessels at sea. The practice was known as CHASE or Cut Holes And Sink 'Em. McCarthy (in USCong 1969c: 50–2) revealed during testimony given to an HFAC hearing that he had been tipped off concerning further plans to transport chemicals (4,000 tons of mustard gas, 2,500 tons of nerve agent, plus tear gas)

[8] Reuss challenged Emerson later in the proceedings after Emerson forgot to bring details of the Army's biowarfare budget with him (US Cong 1969b: 128):

> 'Dr. EMERSON. I have asked someone to get the information for me. I'm sorry. I had it laid out to bring and it was an oversight.
>
> Mr. REUSS. But have whoever is close to your desk pick that up, get into one of those limousines and come on over and tell us. Is that what is being done?
>
> Dr. EMERSON. Not quite.
>
> Mr. REUSS. Why not?
>
> Dr. EMERSON. We don't have a limousine'.

across America by train and then dump them off the New Jersey coastline using surplus World War II Liberty ships. HFAC Committee member, Cornelius Gallagher (D: New Jersey; in USCong 1969c: 2)—whose father had died from gas in World War I—commented on the proposed plan: 'This is a grisly business from the human standpoint. … The train shipment has caused a tremendous concern.' Like the hearings on Dugway, McCarthy's shocking testimony was delivered in May 1969—at the time that NSSM-59 was authorized. Indeed, Laird was already aware of the situation. McCarthy had written directly to the secretary expressing his concerns about the controversial plans. Not unlike other bioweapons-related scandals uncovered by McCarthy, moreover, the incident also figured prominently in the news, engendering a significant public backlash.

CHASE was not the end of the lurid headlines. Details of a chemical accident at a US military base in Okinawa, Japan were released in July 1969. Twenty-three soldiers and an American civilian were exposed to nerve agent when soldiers sandblasting chemical bombs to clean them inadvertently punctured one. Okinawa was quickly followed by equally explosive revelations that the US had secretly stockpiled chemical armaments in West Germany (Fellows 1969: 7; Shuster 1969: 17). West Germany then called publicly for an international ban in response to America's actions. A further nerve agent leak at Dugway in December added fuel to the growing flames ('A Gas Leak', 1969: 33). As one US journalist commented on these incessant and highly public exposés:

> [T]he Pentagon is getting in more and more trouble with the Congress and the civilian leaders of the United States. The trouble is not that the Pentagon is wicked but that it seems to be clumsy: it is constantly being caught out doing things that embarrass the Government and complicate the conduct of American foreign and even internal policy.
>
> **(Reston 1969: E12)**

In addition to these shocking revelations, a number of commercial books critical of bioweapons were published, for example Seymour Hersh's (1968a) *Chemical and Biological Warfare: America's Secret Arsenal* and Robin Clarke's (1968) *The Silent Weapons: The Realities of Chemical and Biological Warfare*. The books helped bring biowarfare even further into the popular public domain and explicitly in a manner that reflected and reinforced taboo perceptions. Biowarfare also extended into fiction. Author Michael Crichton (1995) now published *The Andromeda Strain*—the best-selling story of a gruesome extra-terrestrial disease that kills numerous unsuspecting people in Arizona and almost triggers a nuclear explosion. The novel was published in May 1969. Once again, the date coincides with the authorization of NSSM-59.

Pressure to act

Public and political commentators, even at the time, stated that US policy-makers were unable to ignore the sheer extent of opposition pertaining to these identified scandals ('A Hard Look', 1969: A18; Binder 1978: 18). The situation created a significant pressure to renounce bioweapons. Bennett (in USCong 1969a: 59) said that the events outlined above and the avid resistance that they inspired 'have all converged to mobilize public and congressional demands for reform. I believe that our domestic political climate is more conducive to a revision of our policies on CBW [chemical and biological weapons] than it has ever been before'. Critically, it was the taboo that drove such depth of sentiment. Public opposition to biowarfare was not motivated by calculations of utility or practical consideration. Resistance was stimulated by an intrinsically emotive rejection of biowarfare as disgusting and pariah. For example, Frank Shakespeare (1970: 1)—a later USIA Director after Murrow—sent a memo to Kissinger on the inclusion of toxins within Nixon's renunciation policy in which he cautioned Kissinger on this point. Shakespeare concluded that the non-inclusion of toxins would prove disastrous explicitly due to the 'repugnance with which the public regard such agents' and where that repugnance was 'so great that technical explanations and attempts to justify rationally their possible military use would fall mainly upon deaf ears'.[9] The USIA had produced a briefing paper the previous year highlighting the growing public negativity concerning bioweapons as a taboo matter. The briefing advised that 'psychological factors are a vital element to be considered [with regards to biowarfare policy and policy presentation]' ('Briefing Book', 1969: 4).

Taboo sentiment was intensified by the fetishizing gore and macabre character of the scandals themselves. Dead sheep, death trains, and secret chemical bombs were especially horrific in terms of an established taboo revulsion towards poisonous things. The feeling was exacerbated in these cases due to their seeming uncontrollability: the sheer thousands of animals that were slaughtered; the apparent inability to stop trains laden with murderous chemicals crashing and spreading unfathomable human tragedy; and the lack of

[9] As an example of this public feeling, a *New York Times* editorial on the uncertainty of toxins commented: 'In any event the use of disease-bearing weapons of any sort should be repugnant to—and banned by—every civilized nation' ('... and Toxins', 1969: E14).

control over bioviolence and technology itself, whereby Allen (2007: 154) describes the Dugway sheep as the equivalent of 'Frankenstein's monster'. The public fear that these scandals generated then pressurized, motivated, and was even genuinely felt by, those at the table on NSSM-59. Consequently, public opinion as a conduit of the taboo was key to renunciation. Yet it should also be noted that there remained strong opposition to biowarfare within the military as well, where that rejection was similarly motivated by taboo feeling. Hersh (1968b) said at this time:

> This distaste for C.B.W. [chemical and biological weapons] even pervades parts of the Pentagon; some military men I spoke with conveyed the impression that the use of gases and biological isn't manly; it isn't the kind of warfare that cadets learn about at West Point; it's 'sneaky'.
>
> **(Hersh 1968b: 87)**

Subsequently, Laird (in Kissinger 1969g) concluded that 'biological research is something that can be supported but biological warfare cannot be supported by anyone'.

The US response to public pressure can be portrayed in strategic terms. It may not be the case that a specific decision-maker entirely believed, or came to believe, in the validity of the taboo. Instead, it can be argued that they simply wished to be seen to respond to public resentment on this matter. Doing so would prove beneficial for their own reputation and political standing. A strategic explanation would certainly fit with Tucker's (2002: 128) assessment that Nixon chose to renounce because he wished to be perceived as a 'man of peace' and to 'deflect criticism' pertaining to Vietnam. It would also reflect Laird's belief that a biowarfare prohibition would 'help diffuse the anti-war movement' and remove a key thorn in the US government's side concerning chemical agent use in Southeast Asia (Goldman 2009: 542 fn.40). It has already been established in Chapter 5, however, that the strategic utilization of the taboo does not necessarily negate the norm; indeed, it can reinforce it.

Moreover, the taboo's normative relevance is again vital in explaining how the taboo would offer any value as a strategic resource. Even if it is not accepted that Nixon was genuinely motivated by the taboo, the president would not have drawn upon the biowarfare issue to enhance his reputation and placate an unimpressed public if the taboo did not fetishize the matter as a crucial concern. If the bioweapons taboo had no normative weight, then the renunciation of those armaments would not be associated with peace to such an extent that it possessed political currency as such. Taboo fetishization ascribed these arms a political worth. Elimination would not comprise a meaningful success if bioweapons did not possess normative value. The argument put forward here is especially pertinent as a further contradiction of the utility argument. If bioweapons were unusable then they could not influence political perception as they did not require any sacrifice or concession on the part of an actor. The real value in Nixon's 'man of peace' reputational image was possible only to the extent that he was addressing weapons of extreme import. The importance here was derived from the taboo.

The taboo advocates

The Dugway scandal featured in an NBC television broadcast on 4 February 1969. McCarthy was watching. And he was incensed. Writing later in his book, *The Ultimate Folly*, McCarthy (1970) recalled the evening in question:

> As were millions of other couples across the land, my wife, Gail, and I were sitting in our living room watching television. At nine o'clock NBC-TV presented its 'First Tuesday' magazine-format show. The principal segment documented aspects of British, Canadian, and U.S. CBW [chemical and biological weapons] programs. The Dugway sheep kill incident was pictured. … When my wife fully realized what was being shown on the program, she shooed away our five children, who were curiously peering at the program from the living room doorway, and dispatched them to bed. Shocked by what she saw about U.S. germ and gas warfare projects, Gail peered at me and asked: 'You are a Congressman. What do you know about this?' 'Nothing', I answered, but my interest and indignation climbed as I continued to watch the story unfold; indignation because I realized that I had undoubtedly voted funds for this kind of activity but which apparently were buried in other appropriations bills.
>
> **(McCarthy 1970: 126)**

The television show proved to be McCarthy's turning point. Now convinced that biowarfare was 'utter lunacy' (McCarthy in Raphael 1969b: 3), he swiftly became established as a leading figure opposing the US programme. Meselson (in Conway 1972: 183) said: 'It was like God tapped this guy to publicize this one single issue.' McCarthy was joined on this political matter by other influential actors within Congress. These figures included Senators Charles Goodell (R: New York), Thomas McIntyre (D: New Hampshire), and Gaylord Nelson (D: Wisconsin). McCarthy was exceptionally talented at courting the news media to promote his aims. When the confidential NSC analysis on renouncing toxins was leaked to the press, Kissinger (1969e) instinctively blamed the revelation on 'McCarthy's office' due to the Congressperson's previous record in shaping the headlines through his media contacts. Every scandal that McCarthy uncovered was a newsflash. Critically, McCarthy explicitly sought to promote the taboo via this campaign. He repeatedly stated that his concern was intrinsically motivated by moral feeling and a rejection of biowarfare as pariah. McCarthy (in Hersh 1969: 28) said he believed that bioweapons were 'directly contrary to the principles of humanity and decency for which America stands'.

Fraser (in USCong 1969a: 45) praised McCarthy during a Congressional hearing for the very real impact of his activism. He said: 'I think you have helped changed public opinion, and created an awareness on the part of the public, as well as the Congress, of the urgency of this matter.' As an example of this influence, McCarthy and his allies successfully undermined the bioweapons programme through a Congressional amendment on the 1970 Appropriations bill ('Authorization of Appropriations', 1969). The amendment orchestrated decisive budget cuts to the programme (which removed funding for offensive biowarfare research and development) and introduced staunch new restrictions pertaining to testing and transportation. Those responsible for the amendment articulated this political move as a moral crusade motivated by taboo ideals and expressed through taboo talk. Nelson (1969: 23,073) said that bioweapons could 'only be judged as being demented, because it is the most hideous and debasing form of warfare'. Goodell (in 'Authorization of Appropriations', 1969: 23,192, 23,193) highlighted the taboo themes of excessive destruction and weapons use as uncivilized when he spoke of biological aggression as 'the means by which civilizations of the world could plunge into the abyss of epidemic and mass death' and stated that it posed 'grave moral issues'. The amendment's co-sponsor, Senator Stephen Young (D: Ohio; in 'Authorization of Appropriations', 1969: 23,223), also appealed to disgust when he said: 'Americans are becoming uneasy and concerned about the most grisly weapons in contemporary arsenals—the weapons of chemical and biological warfare.' McIntyre (in Finney 1969a: 4) asserted that any commitment to national security must be 'measured against this nation's traditional opposition to the offensive use of such [biological] agents'. In making such a calculation based on a long-standing and taboo-driven understanding of biowarfare, McIntyre concluded that a major bioweapon programme could not be justified even in terms of research.

The taboo was further evident in the public reception to the amendment. An editorial in *The Christian Science Monitor* commented in response to the resolution that bioweapons were 'regarded as both subtle and vicious. Thus, there exists, deep within humanity's conscience, a built-in readiness to abolish these horrendous devices' ('Destroy this weapon', 1969: 18). Significantly, the amendment was approved midway through NSSM-59 and was highly influential in respect to Laird. Indeed, Laird was zealously vocal in his support for the legislation. He even openly stated that his office had liaised with McIntyre to develop it—to the deep chagrin of the JCS (Weaver 1969). As NSSM-59 progressed, therefore, Laird was working very closely with those fighting for the taboo.

McCarthy and his Congressional associates were not the only taboo advocates. Considerable opposition to the bioweapons programme also existed within the scientific community. For example, the Federation of American Scientists (FAS) condemned bioweapons use in 1965 where this:

> [W]ill hurt our efforts immeasurably in goodwill and moral respect all over the world. We find it morally repugnant that the United States should find itself party to the use of weapons of indiscriminate effect, with principal effectiveness against civilian populations.
>
> **(R. Scott 1965: 11)**

Research collaboration on biowarfare was also heavily contested. The University of Pennsylvania abandoned two secret and highly lucrative military contracts researching biowarfare in November 1966—Operations Spicerack and Summit ('U. of Penn Shuns Germ War Pact', 1966). The decision was made following a student outcry in which the contracts were disparaged as 'immoral, inhuman, illegal, and unbefitting to an academic institution' (Herman and Rutman 1967: 526). The American Society for Microbiology officially ended its advisory association with Fort Detrick in 1968 as it no longer wished to support any research that opposed the 'health and well-being of mankind' ('Tie to Army', 1968: 29). The Smithsonian also became a target for protest after allegations that the institution had been awarded a $2,807,460 US government grant called the Pacific Project to ostensibly study the migratory patterns of birds. The project was said to be merely a cover for secretive research into bioweapon testing (Finney 1969b: 42).

Individual taboo advocates within the scientific community were also key. While McCarthy claimed that he was the one to influence Laird in producing his initial memorandum, others have maintained that this was the work of Meselson (Frank 1975: 108). Known as having an 'unflappable' personality (Rosenbaum 1969: 17)[10], Meselson is a scientist with expertise in genetic

[10] David Rosenbaum (1969: 17) recalled the following example: 'His [Meselson's] unflappable nature was exhibited in a restaurant here a few weeks ago. After he had had a hamburger lunch, the waitress

and molecular biology. He had spent time with the ACDA in the early 1960s, where his experience saw him question the moral basis of biowarfare (Primack and von Hippel 1974: 146). Meselson subsequently became a strong advocate of the taboo. He later described biowarfare as an issue that could 'reach deeply into what we are and how we regard ourselves. It should be evident that any intensive exploitation of biotechnology for hostile purposes could take humanity down a particularly undesirable path' (Meselson 1999: 51). Meselson was already an established campaigner against biowarfare at the time of NSSM-59. He and 21 other scientists, with the support of FAS, had officially written to Johnson in 1966 to protest against biowarfare within the context of Vietnam. The letter explicitly sought 'to warn against any weakening of the world-wide prohibitions and restraints on the use of chemical and biological (CB) weapons' ('22 Scientists', 1966: 3). The FAS circulated the letter within the wider scientific community for more signatures. It eventually became a 5,000-strong petition (Langer 1967a: 302; Langer 1967b: 1073).

Meselson did not personally claim to have influenced Laird. He did, however, claim to have influenced Kissinger. Kissinger was his friend and they had worked together at Harvard University. Meselson said that his involvement with Kissinger on biowarfare was the consequence of a chance meeting between them at Boston's Logan International Airport as each was running to make a flight. The scientist congratulated Kissinger on his new appointment as National Security Advisor. Kissinger replied: 'What shall we do about *your* thing?' The comment was a reference to Meselson's activism against bioweapons and meant that Kissinger was willing to assist on this matter (Klotz and Sylvester 2009: 44). Kissinger would go on to ensure that Meselson's numerous advisory papers against a biowarfare capacity would feature prominently on Nixon's desk throughout NSSM-59—to the extent that Glen Cross and Lynn Klotz (2020: 185) directly attribute Nixon's renunciation to Meselson. Forrest Frank (1975: 108) is more conditional about the scope of Meselson's influence. He believes that Kissinger exploited Meselson to provide both external and scientific validation to a decision that had already been made within the administration. Even if the latter is the case, however, Meselson still provided a crucial sense of credence to the renunciation policy.

Consequently, key advocates were pushing strongly for the bioweapons taboo at this time. The wider public sentiment against these armaments was not only present; it was also successfully facilitated, coordinated, and exacerbated by prominent figures and organizations who sought to uphold taboo

asked him if his meal had been satisfactory. "You might tell the chef," the biologist replied, quietly and without emotion, "that his meat is nearly ready to go bad"."

ideals in doing so. Advocates were also capable of exerting actual political power (e.g. the 1970 budget amendment) and gaining the ear of those in charge (e.g. through advisory opinions and coincidental flight schedules). The advocates both organized public opinion and exerted internal government impact to the extent that their influence actively shaped the decision to eliminate bioweapons. Significantly, this influence was specifically manifest as an expression of the taboo, and the taboo comprised the professed motivation for these actors. As such, the taboo became embedded as a key factor in elimination.

International norms against biowarfare

The UN taboo against biowarfare (outlined throughout the previous chapters) continued to intensify in terms of normative strength and impact at this time. The US was closely involved in this on-going process. For example, it was Eisenhower (1960b) who kick-started what would become the 1967 Outer Space Treaty (OST) with an appeal to the General Assembly that 'no nation will put into orbit or station in outer space weapons of mass destruction' (*Treaty on the Principles Governing the Activities of States in the Exploration and Use of Outer Space, including the Moon and Other Celestial Bodies*, 1967). The discussions leading up to the OST prohibition explicitly encompassed bioweapons within the WMD classification (Bentley 2014: 55). Kennedy was also involved in the OST proceedings. He asserted in vividly taboo terms that any failure to commit to the OST would leave America unable to 'maintain the leadership and respect of the free world … [or] satisfy our own hearts and minds that we had done all that could be done to avoid the holocaust of endless death and destruction' (J. Kennedy 1963: 696). After OST, attention then turned to the oceans and the negotiations for the 1971 Seabed Treaty, which would preclude all forms of WMD from the sea floor (*Treaty on the Prohibition of the Emplacement of Nuclear Weapons and Other Weapons of Mass Destruction on the Seabed and the Ocean Floor and in the Subsoil Thereof*, 1971). The NPT was also signed. Consequently, international eyes were very much on prohibiting extreme weapons including biological arms, as evidenced by the UK draft of the BWC. These various treaties strongly reflected, reinforced, and further institutionalized the stigma surrounding bioweapons that had been progressing within the remit of the UN since 1945. With every associated agreement that came into force, the taboo became increasingly entrenched as a core concept of international political thinking at the UN and the bureaucratic structures underpinning it.

The UN stigma around biological aggression was especially evident in relation to a July 1969 report on biowarfare authorized by Secretary General U Thant. The report was released only six days after the US had publicly admitted the scandal at Okinawa. The document comprised a truly damning account of bioweapons and called for the armaments' immediate and unequivocal proscription. Yet the report did more than simply demand a ban; it did so on the explicit basis that bioweapons were deeply pariah. With emotive echoes of the taboo, U Thant (in UN 1969a: 264) said that he had commissioned the analysis as 'the international community was not sufficiently conscious of the dangers inherent in this new type of *weapon of mass murder*' [emphasis added]. The primary content of the report was still a practical assessment of weapons capability as U Thant had wanted the report to act as a means of improving the general understanding of these armaments and what they could do. The analysis was still, however, effectively a mandate for the taboo and a rejection of bioweapons due to disgust and repulsion. Biological armaments were dismissed as a 'universal horror', causing 'grave dangers', and that 'must be clearly classified as weapons of mass destruction' due to their intolerable and unethical effects (UN 1969a: 264, 266, 290). The report additionally drew on historical perceptions of biowarfare as morally forbidden in order to demonstrate the inherent and long-standing repugnance pertaining to these devices:

> No form of warfare has been more condemned than has the use of this category of weapons. The poisoning of wells has been regarded from time immemorial as a crime incompatible with the rules of war. 'War is waged with weapons, not with poison' ... declared the Roman jurists.
>
> **(UN 1969a: 267–68)**

To the extent that the report did still refer to military utility claims it should be noted that one of the authors was Bennett. It has already been demonstrated in this chapter that Bennett saw a need to employ practical claims in order to win the taboo argument. Focusing heavily on the military detail, therefore, did not negate the influence of the taboo in this case or suggest that the report did not express a commitment to normative feeling.

The US explicitly recognized that the taboo comprised a key international norm in this respect. Officials repeatedly acknowledged that the global normative stigmatization opposing biological aggression exerted a significant pressure on them to proscribe these armaments in some meaningful way (e.g. ACDA 1969: 827; G. Smith 1969: 4–5). Kissinger (1969h) penned Nixon a memo stating that the negotiation of the NPT and Strategic Arms Limitations

Talks (SALT) meant that the demands on America to clarify its position on biowarfare would now multiply. The international community would expect that the prohibition of nuclear arms would be extended to bioweapons in that these were also categorized as WMD. Samuel De Palma (1969: 375), Assistant Secretary of State for International Organization Affairs, further commented that bioweapons were set to 'emerge as the next important area for arms control negotiations'. Kissinger (in 'Minutes of a Review Group Meeting', 1969: 326) elsewhere noted that bioweapons could no longer be considered a low political concern. He asserted that the extra attention now being paid to biological violence could not be ignored as 'high level interest sometimes brings with it higher priorities'. Consequently, both internationally and domestically, the taboo now comprised a compelling expectation that biowarfare should be eliminated as a priority and precluded as a pariah form of weaponry.

Transparency and public opinion

Renunciation has previously been reduced to a simple calculation of military utility. If bioweapons were strategically and tactically irrelevant, then why retain them? A conclusion based purely on practical calculations of weapons use, however, overlooks the following factors relating to the taboo: evidence of normative pressure manifest in taboo talk and public scandal; genuine or reputational commitments to the taboo expressed by key decision-makers; the adoption of an outright ban that went beyond the scope of hard-line military logic; the sheer potency and groundswell of wider feeling at both the international and domestic levels which upheld the taboo belief that bioweapons are pariah; and the coordination of public and political opposition by persuasive taboo advocates.

The focus on the taboo here is not to state that perceptions of practical military advantage were entirely irrelevant in respect of renunciation. In particular, an actor's belief that they were not sacrificing a critical asset in abandoning bioweaponry would make renunciation easier even if that decision were normatively motivated. Yet the taboo still played a decisive role. As Paul Doty, Mallinckrodt Professor of Biochemistry at Harvard, concluded during a November 1969 meeting of the American Academy of Arts and Science on chemical and biowarfare: '[T]he direction of United States policy will be determined as much by the *public conscience* as by any technical military considerations' [emphasis added] ('Chemical and Biological Warfare', 1970: 14). Indeed, engaging the public moral feeling was seen to comprise a fundamental aspect in promoting the elimination of bioweapons. The 1969 UN report stated that its publication was explicitly intended 'to contribute to public awareness of the profoundly dangerous results if these weapons were ever used' explicitly in order to then place pressure on governments to abandon biological armaments in line with the taboo (UN 1969a: 267). The report authors believed that public support for the taboo represented the key means by which normative expectations could be realized at the political level.

The preceding chapters already demonstrate that public opinion had long functioned as a catalyst with respect to the taboo. In relation to renunciation, however, an even deeper awareness had developed concerning what the US bioweapon programme entailed and meant. The public, and even key political actors such as McCarthy, had previously known relatively little concerning the military facts of the programme. McCarthy (1970: 123) described the situation as one in which biowarfare policy was obscured by 'the totally unnecessary curtain of secrecy that covers even the most pedestrian aspects of CBW [chemical and biological weapons]'. Secrecy was attributed to the taboo and the associated fear of negative backlash should the norm be seen to have been contravened. McCarthy (1970: 61) further commented: '[T]he U.S. Army's biological warfare experts are reluctant to discuss the morality of their weapons.' Elinor Langer (in 'U.S. Said to Study Disease', 1967: 10) similarly argued that the silence surrounding the bioweapons programme was maintained as a means of precluding negative public opinion. Langer explicitly stated that public opinion was the decisive issue in this matter as opposed to more military-driven explanations such as protecting national security.

McCarthy (1970: 123) additionally stated that the clandestine approach to biowarfare had contributed to a 'failure to subject CBW [chemical and biological weapons] to rigorous policy analysis following World War II'. The taboo could not be comprehended within the usual means of democratic contestation because the facts had not been in the public domain and therefore could not be questioned. Secrecy had then facilitated both the military's control over the programme and had also limited any input by other actors who may have possessed a more positive appreciation of the taboo. One anonymous microbiologist at Detrick claimed that the programme's covert nature had caused it to become 'dominated by military thinking … not only that, but the dregs of the military' (in Hanrahan 1969: A20). The military had not been sufficiently challenged or held to account. These actors could then pursue aggressive biowarfare policies that might have otherwise proved untenable if subject to wider public scrutiny.

Yet the bureaucratic mystery surrounding the bioweapons programme had now broken down. A *New York Times* editorial welcomed NSSM-59 specifically as a strike against the secrecy that had clouded the bioweapons programme. The article argued that the programme had been concealed to a degree that undermined the need for that strategy's own justification. NSSM-59 offered a chance to change that approach: 'Congress can take down the "Please Do Not Disturb" sign from this program and begin to discharge its normal function of review and debate' ('War With Gas and Germ', 1969: 42).[11] Both politically (in terms of NSSM-59 and McCarthy's Congressional campaign) and more publicly (in terms of the shocking scandals of dead sheep and the potentiality of indiscriminate killing in Vietnam) bioweapons were now more widely acknowledged and understood. Opening the programme up to this public review shed a new light on the US biowarfare capacity that had previously been denied. And greater awareness demanded greater action with respect to bioweapons elimination. McCarthy (1970: vii–viii) said of this matter: 'Yet when the facts have been made public, Americans have made their opposition to these [pro-biowarfare] policy changes known.' Consequently, the focus here on what was known reveals that renunciation was not simply a question of public opinion. The policy was also specifically an issue of public transparency and what information had been communicated to other actors.

Public opposition was exacerbated by a loss of trust in the US government and military institutions concerning biowarfare as a further consequence of this new information. Revelations concerning the lack of transparency surrounding US possession of bioweapons and the scandals that the programme had produced tarnished public faith. Science Advisor, Lee DuBridge (1969: 2), commented in a top secret memorandum to Kissinger during NSSM-59: 'There is a large reservoir of scepticism, cynicism, and incredulity that has developed as a result of our past lack of policy and the inconsistency of past statements in this area.' In particular, officials were not trusted due to military actors actively lying about bioweapons; for example, the initial denial of what had happened at Dugway ('Candor', 1969: 46; Boffey 1968).

Ultimately, public opinion bolstered the taboo in respect of renunciation. The case study here has demonstrated how a taboo can become strengthened when any cover-up of its violation fails and a full(er) knowledge of the norm's transgression or potential transgression comes to the fore. Once it was publicized that the taboo was under threat of violation, public actors sought to reinforce that normative expectation. As the programme became a more widely known issue, so the taboo increasingly stymied activities that did not

[11] Elsewhere in the *New York Times* biowarfare research was described as the DoD's 'dark corner' ('Defense's Dark Corner', 1969: 46).

comply with a normative rejection of biological armaments. Critically then, the taboo also comprised a core explanation for the call for renunciation. Lincoln Bloomfield (in 'Chemical and Biological Warfare', 1970), Professor of Political Science at MIT, spoke at the same American Academy meeting mentioned previously. He said that the public fear and repugnance associated with bioweapons comprised the most powerful argument of all against their use:

> There is one argument that transcends all the technical, military, strategic, and economic pros and cons—that is the psychological argument. . . . From what I have read and learned and heard now at this meeting, I believe the most important thing to be done is not technical, not legal, but psychological. It is to reinforce the expectation that CW/BW [chemical weapons/biological weapons] will not be used; it is to reinforce the inhibition against it.
>
> **(Bloomfield in 'Chemical and Biological Warfare', 1970: 13)**

The taboo was also reflected within the decision to renounce itself. Senator Frank Church (D: Idaho; in USCong 1975: 126) highlighted the unilateral nature of renunciation at a select committee hearing in 1975. He said that the strength of this commitment by America against biowarfare was 'a gesture for decency and a gesture for peace'. Renunciation, therefore, was not an expression of practical utility designed to convince others of the same logic. Church understood that the US's abandonment of any biological armaments capacity was a moral decision. It was the 'right' thing to do. Consequently, the bioweapons taboo was now upheld as a more robust normative standard and an established public concern. In particular, renunciation was manifest as a major step forward in respect of codified stigmatization at the US level. The rejection of biowarfare was now expressed in domestic law. The taboo was now not merely shaping but actively *deciding* the political agenda. Moreover, the commitment to the taboo was about to culminate not only in the policy of US renunciation, but an even stronger and genuinely international realization of the norm: the BWC. This agreement is the focus of Chapter 7.

7
Prohibition
Biological Weapons Convention

> [R]epugnant to the conscience of mankind.
>
> **(Biological Weapons Convention, 1972)**

What is the first thing that you would do if you no longer possessed a certain weapon? Ensure that no one else possesses it either. Why put yourself at risk by allowing others to have what you do not? Consequently, self-interest has been cited as the primary reason for America's post-renunciation move to advocate for the BWC and to ratify the Geneva Protocol (Beard 2007: 280). The US was willing to pursue all means of global elimination once they had decided to abandon bioweapons for themselves.

Yet an exclusively self-interested explanation again overlooks the cardinal relevance of the bioweapons taboo. The reference here to the taboo is not limited to the repugnance felt towards bioweapons explicitly, but also relates to the international norm to prohibit all WMD manifest at the UN. The preceding chapters document the intensifying stigmatization of WMD and the institutionalization of that normative expectation within UN diplomatic structures. While the decision to prohibit bioweapons can still be linked to self-interested impulses in part (discussed further later in this chapter), the previous pages have demonstrated that the BWC did not emerge from nowhere. The agreement was negotiated as part of an extensive and prolonged process of normative development. The identified trend connects in particular to the negotiation of the NPT. The successful conclusion of a global treaty on nuclear weapons not only realized a prominent commitment to remove these harrowing devices. The NPT also opened arms control discourse up in relation to other weapons classified as of major concern, per the collective nature of the WMD concept. It was argued that if consensus could be achieved on nuclear arms then surely similar agreements could be enforced on comparable forms of stigmatized weaponry. Expanding prohibition to all

The Biological Weapons Taboo. Michelle Bentley, Oxford University Press. © Michelle Bentley (2023).
DOI: 10.1093/oso/9780198892151.003.0008

mass destructive technologies had always been the stated intention (Bentley 2014: 59–60). Eliminating bioweapons was, therefore, the next step.

Consequently, the BWC marked the official and substantive realization of a major global norm. The bioweapons taboo had existed for centuries and had already undergone significant codification, such as the Geneva Protocol. Yet the BWC still constitutes the most critical juncture within that history. The BWC was the moment at which the norm became fully established within a dedicated treaty devoted to upholding the taboo as an integral aspect of the international community. The agreement was also the first treaty ever to prohibit—to use the words of then US Ambassador to the UN and future President, George H.W. Bush (1972: 104)—an 'entire class of weapons'. Adherence to this prohibition may have been, and remains, far from perfect. It was through the BWC, however, that the taboo truly came into its own.

The chemical problem

It was not the case that US renunciation led seamlessly to the global prohibition of bioweapons. The core barrier to signing on the BWC's dotted line, however, was not biowarfare itself. The main problem was chemical weapons. Normative rejection of chemical aggression had facilitated the push towards renunciation. Yet continued American deployment of chemical agents in Vietnam would ultimately threaten the BWC process and the ratification of the Protocol within the US. Indeed, renunciation actually served to negatively exacerbate the situation regarding chemical use in the Vietnam conflict. NSC staffer Michael Guhin (1970) sent a note to Kissinger summarizing a memorandum by DuBridge. Guhin commented that, while the domestic and international response to bioweapons renunciation was 'generally very favourable', the announcement had inadvertently also 'touched off another round of sharp criticism of our use of tear gas and herbicides in Vietnam'.

Central to this criticism was an accusation that chemical employment in Southeast Asia violated the terms of the Protocol. US officials still steadfastly maintained that this was not the case. It was repeatedly stated that the Protocol did not cover non-lethal and incapacitating agents such as riot control and defoliants. There was no violation; more specifically, there was no violation of the taboo. Yet the dispute on this matter triggered significant tension at the UN. For example, Hungary introduced a resolution in 1966 calling for a universal adherence to the Protocol. The proposal was couched in highly undiplomatic language and constituted an ill-disguised attack on the

US concerning Vietnam.[1] ACDA Director, William Foster (1966b: 759–60), responded with indignation that America took 'exception to the nature and tone of the remarks' of the proposal, which included 'distorted charges and deplorable innuendos' (see also Nabrit 1966: 800). The resolution was refuted by the US as yet further Communist propaganda and a malicious attempt to exploit the Protocol in order to disparage American foreign policy (Sisco 1967: 459).

Similar tension was evoked concerning a further resolution put forward in December 1969—not by a Communist faction this time, but by Sweden. The document also demanded an official recognition of the Protocol's customary scope. The text expressed the taboo sentiments that 'chemical and biological methods of warfare have always been viewed with horror and been justly condemned by the international community' whereby their effects were 'inherently reprehensible' (UN 1969b). The resolution was approved 80–3 with 36 abstentions. The US was one of the three. American officials later justified their negative vote as a symbolic statement that the General Assembly was not the appropriate forum to mediate treaty interpretations or procedure. Yet they were also keen to highlight that the US vote did not comprise a direct rejection of the Protocol itself or a breaking of the taboo (USCong 1971: 17). The US delegation maintained that America upheld the taboo principles inherent in the Protocol. The decision to reject the resolution was instead a fundamentally bureaucratic concern. Despite this explanation, however, and while the US continued to protest that it had done nothing wrong with respect to chemical agent use in Vietnam, this issue became a persistent source of international backlash that could not be shaken off.

The controversy would play into the BWC. The UK submitted a draft version of the agreement to the Conference of the Committee on Disarmament (CCD) on 10 July 1969. The draft was notable for what was not included: chemical weapons. Many actors had assumed that further arms control in this field would address both chemical and biowarfare concurrently, as the Protocol had established a precedent in doing so. The UK countered, however, that key differences between the two forms of weaponry necessitated independent prohibitions. Separating the two types of armament constituted the most pragmatic option at the very least. British Disarmament Minister, Fred Mulley (1968: 537), explained the UK view that chemical weapons had been employed in conflict whereas bioweapons had not. It was simply easier to proscribe a weapon prior to its usage as opposed to clawing back one

[1] The final document contained clear reference to the taboo, whereby it stated: 'Considering that weapons of mass destruction constitute a danger to all mankind and are incompatible with the accepted norms of civilization' (UN 1966).

already actively assimilated into military strategic thinking and action. American actors including Leonard repeated Mulley's line of argument (Bentley 2014: 59–63). Bioweapons prohibition would be more achievable, and achievable more quickly, than a multifarious ban that must also address the distinct complexities of chemical violence.

The decision to separate biological and chemical weaponry was elsewhere blamed on the on-going debacle of Vietnam. The UK was perceived as indulging the US. Including chemical weapons within the scope of the draft would have proven extremely awkward for the latter. Highlighting chemical weapons within a major disarmament document would intensify the pressure to cease chemical deployment in Southeast Asia and cause the US a significant humiliation. The CIA (1969a) said:

> The British government has remained sensitive, however, to the fact that the US considers tear gases and other non-lethal substances permissible under the Geneva Protocol and international custom. Deference to the US no doubt played a part in Britain's decision not to treat CW [chemical weapons] in its present initiative [the BWC draft], since debate on the tear gas issue, part and parcel of any CW discussions, is potentially embarrassing to the US. London has adopted the 'piecemeal' method of splitting off a subject upon which agreement is possible while ignoring a potentially volatile issue.
>
> **(CIA 1969a: 14)**

Critically, British deference did not equate to the approval of US activity in Vietnam. Joshua Lederberg (in USCong 1969a) was a molecular biologist and a taboo advocate within the scientific community.[2] He told a Congressional hearing:

> The British have bent over backward not to say anything nasty to us, because they like to think of themselves as our friends, but officials in the British Government really feel very strongly that the way we have been using tear gas in Vietnam simply constitutes chemical warfare: that it is wrong.
>
> **(Lederberg in USCong 1969a: 126)**

Lederberg's comments perhaps overlook the hypocrisy of the UK stance on riot agents, given that the British government was content to employ such tactics in Northern Ireland. Yet the BWC draft was still interpreted as a concession to US interests.

[2] The *New York Times* reported on Lederberg's previous opposition to bioweapons and his lobbying of the US government on this issue ('Scientist Criticizes Germ War Research', 1969: 32).

Concession, however, served to stoke the controversy that it sought to avoid. The USSR retaliated with a draft of their own to proscribe both biological and chemical aggression. The document fuelled international dissent on this point and threatened to scupper the US position (CIA 1969a: 17; Leonard in USCong 1975: 127–8). Foster (1969: 60) did publicly welcome the ethos of the Soviet submission as an expression of the taboo. He stated that the effort to eliminate taboo armaments 'properly reflects the serious concern which has been expressed over the potential threat to mankind posed by the development and possible use of lethal chemical and biological weapons.' Yet Foster also rejected the dual approach that was proposed. It can also be noted here that Foster's approval was explicitly limited to 'lethal' agents only. As such, his comments supporting a ban implied the exclusion of riot agent and herbicides in the same way as the US line.

Resolution and resolution

Stalemate seemed inevitable. The CIA (1970) cautioned that the controversy surrounding a dual prohibition would constitute a deal-breaker. Agents had received intelligence of a conversation between an unnamed USSR official and a US delegate in the General Assembly during which the former had stated 'that the subject [BWC] has a "high political content" in Moscow and, for that reason, the Soviets are unlikely to support a separate ban.' The Soviet stance was again reduced to propaganda. The CIA described the official's comment as 'a rare admission that Moscow is motivated by propaganda aspects of the CBW [chemical and biological weapons] issue' ((CIA 1970)). At the same time, however, the Soviets were preparing to U-turn. In a confidential telegram sent two months prior to the above CIA analysis, Secretary of State, William Rogers (1970: 2), revealed details of a 'long informal luncheon' between US and Soviet CCD delegates. Deputy Representative for the Soviet side, I.I. Tcheprov, was asked whether the USSR could consider a bioweapons-only prohibition. He replied: '[Y]ou already have it.' The attendees then proceeded to coordinate how the US should make private representations to the Soviet delegation outlining their case. Tcheprov pledged: 'I will do my best [to promote a bioweapons agreement].'

Rogers (1970: 2–3) also referred to further intelligence from the Polish CCD representative, K. Zybylski. Rogers speculated on the basis of this information that the Soviets were now seeking to moderate their position in response to the SALT negotiations. He believed that the USSR had reasoned 'that chances of agreement in Vienna [SALT] would be enhanced if they demonstrated serious and flexible attitude toward disarmament matters under discussion in Geneva [the BWC]'. A CIA (1971a: 4) bulletin later expressed similar conclusions. Soviet Ambassador, Alexei Roschin, had informed Leonard on 29 March 1971 that the USSR was to submit a new draft addressing bioweapons alone. The CIA analysis noted that this shift in policy coincided with the opening of the 24th Soviet Party Congress. The CIA suggested 'that the Soviets may try during the speech-making to portray it [a revised draft] as fresh evidence of Moscow's good faith and flexibility in pursuing limited disarmament measures'. The prediction rang true. General Secretary, Leonid Brezhnev, mentioned the exclusively biological agreement in his keynote address explicitly as a means to highlight improved relations with the US on disarmament ('Brezhnev Speaks', 1971: 44).

Other self-interested explanations for the swing in USSR policy were also put forward. US officials argued that the Soviets had felt that propaganda relating to biological and chemical armaments would not prove as influential once America had withdrawn from Vietnam—hence the shift away from the issue ('Soviet move', 1971: 3). Elsewhere, it was claimed that prohibiting bioweapons only would facilitate opportunities for USSR chemical use against the Chinese, where it was additionally alleged that the Soviets already had chemical arms stockpiled on their border with China ('US urges Russia', 1971: 4). Mangold and Goldberg (1999: 62) further suggest that the concession was merely a ploy to force international bioweapons disarmament, leaving the USSR free to clandestinely develop *Biopreparat* to their advantage. Yet the Soviet decision to abandon a dual approach was still considered to reflect a genuine desire to get something done on bioweapons prohibition, even if this necessitated a yield to the US position. Leonard (in USCong 1975) said:

> When they did agree to this, it was basically as a concession to us because they wanted to move the whole question of disarmament forward, and if this was the way we preferred to do it, well and good, let's do it that way.
>
> **(Leonard in USCong 1975: 127–8)**

Rogers (in Kissinger 1971: 4) similarly commented that it was 'particularly noteworthy that the Soviets have shifted from a long standing position with propaganda appeal to serious negotiations with the U.S.' Consequently, Soviet preparedness to do so was interpreted as a sign of authentic intent on the Communist part. Officials believed that the USSR was motivated by concerns for the taboo on this matter. A CIA (1969b) report stated: 'We believe that political considerations would weigh heavily against Soviet initiative of BW.

In Soviet writings the subject is linked with nuclear and chemical warfare in terms that indicate a high degree of political control and restraint.'

If the USSR had conceded a propaganda tool by abandoning the dual draft, the Soviets now proceeded to make up for it with another tactic. Soviet officials utilized the situation to construct an image of gracious compromise. The USSR portrayed itself as having been the bigger party in breaking the diplomatic deadlock by agreeing to a bioweapons-only agreement (Lusinchi 1971: 1). US irresponsibility had caused the impasse; Soviet goodwill had resolved it. The USSR's positive public presentation was, however, somewhat knocked in May 1971. An official admission of a Soviet bioweapons programme was made after the state had previously refused to confess that any programme existed (CIA 1971b). Deputy Chief of the Soviet delegation to the CCD, V.V. Shustov, had asked a US delegate during conference proceedings whether the US would consent to a clause in the BWC requiring a formal confirmation that American bioweapon stockpiles had been destroyed. The US representative inquired if the Soviets would be prepared to do the same in return. In doing so, the USSR was effectively being asked whether there were stockpiles for them to report the destruction of. Shustov verified that biowarfare arsenals existed. The US had essentially already known that the Soviets maintained a biowarfare capacity (although a CIA analysis highlighted that Shustov's statement remained unconfirmed in terms of the physical evidence; CIA 1971b). The incident, however, was still politically awkward in that it was a public affirmation of American concerns. Yet the admission failed to stymie wider efforts to bring the BWC about. On 5 August 1971, both the US and USSR submitted identical, but separate, texts of the proposed convention. The final version was received on 28 September. Expressing the taboo, the resolution text forbade bioweapons on the basis that these armaments were 'repugnant to the conscience of mankind'. The UN voted in favour of the BWC 110–0 (France abstained) on 25 February 1972.

'We've got to say these things'

US attention now turned to a signing ceremony. Rogers wished to make it clear, however, that the BWC signing was not deemed as significant as that of the NPT. The Secretary proposed that the ceremony not be held at the White House, where the NPT had been signed. The signing should instead take place in the International Conference Room at DoS. The latter venue had been used for the signing of the Seabed Treaty. Rogers (1971: 2) commented: 'While the BW [biological weapons] Convention is unquestionably a noteworthy achievement, it does not rank in importance with the NPT and a ceremony in the State Department would therefore seem to be the most appropriate.'

White House or not, Nixon initially refused to show up at all. A consternated Rogers (in Guhin 1972a: 1) wrote to the President on 9 March 1972, imploring Nixon to reconsider on the grounds that: there existed precedent in terms of his presence in signing the NPT and Seabed Treaty; the Soviets had made it clear there should be 'no appearance of downgrading' of the BWC compared to these other treaties; the Soviets would construe non-attendance as a 'snub'[3]; non-attendance would additionally offend the British, not least as the UK Prime Minister was planning to participate in a respective ceremony in London; Nixon had been an integral figure regarding renunciation and this policy had comprised a key aspect of the wider BWC process; and non-attendance would be interpreted as giving two fingers to China who had not been supportive of the BWC, with potentially negative diplomatic consequences down the line. Chief of Staff, Harry 'Bob' Haldeman (1972), called the President on 17 March to convince him to participate. Nixon finally conceded: 'All right, I'll do it.'

But Nixon was angry. He felt that the Soviets had abused the BWC process to repeatedly censure America in respect of Southeast Asia. The war itself had also recently escalated thanks to USSR-backed offensives by the Democratic Republic of Vietnam. Nixon did show up at the signing on 10 April 1972, and he would strongly assert that the BWC had 'great significance'. Yet he primarily used his speech to slam the Soviets (Nixon 1972a). Nixon (in 'Editorial Note', 1972) said in meetings after the ceremony:

> I'll bet ya' that the disarmament boys at ACDA are probably just shitting their pants because of this thing [Nixon's speech] ... because we should have kept the emphasis on peace and all that. The hell with them. That was senseless to make us sit there with the Soviet Ambassador at a time when they're raping South Vietnam and say that they have made a great contribution to peace by signing this silly biological warfare thing which doesn't mean anything.

> I've heard all the crap about, you know, that science—that we'll work in peace. We all want that. Everybody believes in that. ... But, for an opportunity like this to be missed. For the Soviet Ambassador sitting there ... and now at the present time

[3] Rogers' stance reflected concerns raised by ACDA Director, Gerald Smith (1971), the previous year. Smith said that the Soviets had made numerous concessions in abandoning their commitment to a dual prohibition and, as a consequence, they would consider themselves insulted if the President did not show full and unequivocal support for the BWC, as he had done on other WMD-related agreements.

> with Soviet tanks and Soviet guns rumbling through South Vietnam and for me to sit there and sign it say, 'Isn't it great that we've now had three agreements and we hope to have a SALT agreement'. I couldn't do it. So I just said, 'By God'. Don't you agree? We've got to say these things.
>
> **(Nixon in 'Editorial Note', 1972)**

Nixon's comments ostensibly display little regard for the BWC (the 'silly biological thing'). His expression, however, should not be interpreted as a direct dismissal of the agreement or of the bioweapons taboo. Nixon clearly sought to exploit the BWC signing for political gain and acrimonious point-scoring. Yet Nixon did acknowledge the prohibition's significance in terms of peace and peace-creation as expressed in taboo terms. Indeed, the taboo was again evident in Nixon's (1972b: 455) subsequent letter of transmittal to Congress of the 11th annual ACDA report in which he said of US support for the BWC: 'The presence of these [bio]weapons in the arsenals of any civilized nation is no longer justified.' Nixon's comments at the signing also demonstrate that the President believed this taboo sentiment to be self-evident. The taboo as connected to ideas of peace was so obvious that it did not require direct or extensive expression ('[w]e all want that'). Consequently, Nixon's criticism of the BWC did not comprise a rejection of the agreement or the normative ideals underpinning its provision. His was a concern over how that agreement was being employed politically.

Ratification

There was an additional snag: the US had signed the BWC, but not ratified the agreement. There was no clear objection in principle to doing so. Yet the issue would once again become embroiled within the controversy of Vietnam as well as the US ratification of the Geneva Protocol.

Nixon made a commitment to submit the Protocol for ratification when he announced his policy of renunciation. It took him nine months to do so. The agreement was only transmitted to the SFRC on 17 August 1970. A memorandum by Kissinger (1970) attributes the delay to concerns that the process would become mired in the Vietnam dispute. Paradoxically, however, that same controversy later became the very reason to resubmit. Kissinger (1970) recognized that any further delay beyond the next General Assembly session (commencing in the September) would potentially expose the US to accusations of prevarication and hypocrisy relating to Vietnam, causing 'considerable embarrassment'. Kissinger was also aware that backing for ratification had increased significantly within the Senate. As Senate Democratic leader, Mike Mansfield (D-Montana; in Naughton 1969: 3), told news reporters: 'I see no reason why there should be any controversy [concerning ratification]. It's 44 years overdue.' Kissinger reasoned that Senate support for approval might see it prepared to simply overlook the Vietnam controversy and continuing arguments over American use of chemical agents in favour of ratifying the Protocol immediately. Furthermore, the US was now the only major state that had not ratified the Protocol (Finney 1971: 13).[4] It was consequently deemed unfeasible for the US to maintain a position outside the Protocol and that non-ratification 'could be used for political or propaganda purposes against us' (Guhin 1972b: 1).

Yet the devil was in the detail. The Protocol was reintroduced to SFRC on the basis that non-lethal chemical agents would not be included. The exclusion was designed to placate the DoS and ACDA (Finney 1971: 13). The SFRC rejected this interpretation as unworkable and undesirable. It was argued that international understanding of the Protocol's scope clearly proscribed such chemicals and that any deviation from this viewpoint by the US would prove problematic. The situation descended into an acrimonious standoff between the SFRC and the White House. Long-standing SFRC Chair, James William Fulbright (1971a), reported on 28 June 1971, that the Protocol was now 'deadlocked'. Fulbright still believed ratification to be (in words of the taboo) an 'action [that] would be regarded as truly courageous and possessed of real moral force'. Yet repeated efforts to resolve the situation failed. For example, the SFRC complained that it had written an official letter to Nixon in an attempt to initiate a solution and had not even received a reply (Fulbright 1971b).

Under the impression that the administration was prepared to let the Protocol flounder, the SFRC upped the stakes. The SFRC threatened to suspend the BWC's ratification[5] if Nixon did not concede on the Protocol (Guhin 1972b: 1). The BWC required its three depositories—the UK, USSR, and US—to ratify before the agreement could come into official force. Consequently, any delay on America's part would not only jeopardize a key treaty but also inflict a severe international humiliation on the US itself (Guhin 1974). The US could not be seen to hold up a major global agreement simply because it was incapable of controlling its own Senate.

Failure to ratify the Protocol fast became 'a constant source of embarrassment to this country'—as expressed by the future Chair of the HFAC,

[4] The US and Japan had been the two major state holdouts. Japan ratified on 21 May 1970.

[5] The BWC was submitted to the committee on 10 August 1972.

Clement Zablocki (in USCong 1974a: 6). Furthermore, the US authorized the Paris Peace Accords in 1973. America's withdrawal from Vietnam helped take chemical usage off the table in terms of both the conflict and the associated scandal that had dogged US credibility. Nixon also now announced the CC's disestablishment. Removing the CC undermined the fight to keep non-lethal agents, as well as any outstanding support for bioweapons more generally, by eliminating a key lobbying force (D.S. 1973: 664). The same fight additionally lost other major players such as the American Chemical Society. The organization publicly reversed its position on the Protocol reservation concerning chemical agents and refused to support the administration if it persisted in efforts to apply it (Finney 1973: 7). Moreover, the UK and USSR had now both approved the BWC, leaving America as the only barrier to the Convention.

In what was building to be something of a perfect storm, President Gerald Ford's administration instigated a new push to approve the Protocol in 1974. Ratification centred on a proposed compromise. The US would renounce the first-use of all chemicals but would make an informal exception in the following circumstances: defoliant employment to clear the ground around the defensive perimeter of military bases and installations; defensive riot agents in cases where this use could save lives; and use in certain limited circumstances, for example controlling riots among PoWs and on rescue missions. All usage would require prior Presidential approval. ACDA Director, Fred Ikle (1974: 339), outlined the compromise to the SFRC at a hearing on 10 December 1974. Ikle also emphatically stressed the need to ratify the BWC. Ikle outlined three core reasons to approve the Convention: bioweapons' lack of military utility; 'biological weapons are particularly repugnant from a moral point of view'; and BWC adherence would deter a biological arms race. Consequently, the taboo constituted a major aspect of his argument.

The SFRC were not impressed by the Ford compromise. The proposal was met with dissatisfaction and frustration; but also, with pragmatism. While top SFRC representatives felt that the suggestion was far from ideal, they accepted that it was also the best offer they would receive (Finney 1974a: 12). Moreover, Fulbright had now been outvoted in the 1974 Democratic primary election by then Governor Dale Bumpers (Reed 1974: 85). Fulbright was keen to tidy up his legislative board before leaving as SFRC Chair. There subsequently ensued a 'surprising' and secret compromise between the SFRC and the administration (Shapley 1975: 149). On 12 December 1974, it was finally agreed to let the Protocol pass without a formal reservation, but with the informal interpretation applied (Finney 1974b: 1). The agreement was sent to the Senate floor and unanimously approved four days later.

Ford signed the instruments of ratification for both the BWC and Protocol on 22 January 1975. The signing was carried out at the White House. Given Rogers' previous comments on the BWC and where it should be signed, the ceremony's location highlights the sheer significance attributed to the act. The President said at the event:

> Finally, I believe that these acts of ratification demonstrate the desire of our nation to create and to contribute to a more peaceful world, and I pledge to you that I will continue in the search for new measures to promote that cause.
>
> **(Ford in 'Editorial Note' 1975: 379)**

The BWC came into force on 26 March 1975. In 1976, *The Law of Land Warfare* (which had comprised such a marker in establishing first-use in 1956) was once again updated to incorporate the BWC and Protocol restrictions. The document now acknowledged that biowarfare 'has been justly condemned by the general opinion of the civilized world' (DoA 1976).

Endorsing the taboo

The confirmation, stigmatization, and codification of the taboo via the BWC was more than once threatened by concerns related to Vietnam. Yet the critical point is that it was chemical—not biological—arms that were problematic in terms of the BWC's negotiation and ratification. The greater bearing of chemical weapons is unsurprising. The employment of chemical agents in Vietnam was an extremely prominent issue on both the domestic and global political agendas. Moreover, the bioweapons question was already effectively resolved as a consequence of renunciation. Bioweapons were now banned from the US perspective. Consequently, biowarfare was not a point of major controversy compared to the actual use of chemicals in Southeast Asia.

A parallel can be drawn between Vietnam and the debate on unconventional warfare following World War I discussed in Chapter 3. Chemical arms had been used in the trenches of the World War whereas biowarfare still remained a fundamentally hypothetical threat. The lived experience of chemical weaponry caused the issue to overshadow a discourse that also addressed biological modes of violence. Chemical agent employment in Vietnam similarly shaped the debate concerning the BWC. Critically, the construction of debate on these terms does not negate the relevance of bioweapons and the associated taboo. The nature of that dialogue explains, however, why chemical arms could disrupt political discussions that pertained to bioweapons explicitly. Furthermore, the relevance of the chemical issue demonstrates how the BWC could be contested but that this disruption cannot be attributed to a lack of commitment to the bioweapons taboo. Dispute surrounding the BWC was not motivated by a rejection of the biowarfare norm. Any barrier to codifying the taboo through the BWC was indirectly related to other concerns and the taboo itself was upheld.

Admittedly, the impact of Vietnam indicates certain limitations on the taboo. The taboo was vulnerable to both self-interest and factors apparently deemed of higher priority, such as chemical use. The relevance of these other issues suggests that there was at least a conditional observance of the bioweapons taboo in that the norm was not sufficiently absolute that it came before all other concerns.[6] Yet the taboo was still robust to a significant degree. Ratification was a politically convoluted process and bioweapons did become a pawn in the Vietnam chess game. The actors involved in the discourse on the BWC, however, did not question the Convention on the basis that a biowarfare prohibition was improper or undesirable. The taboo's validity was never doubted or put at stake. While actors claimed to be willing to effectively sacrifice the taboo in order to gain political ground on the Vietnam issue (such as the SFRC being prepared to delay the BWC's ratification due to the Protocol's interpretation of chemical arms), this apparent conditionality did not comprise a direct negation of the taboo. The threat to abandon the BWC was simply a political tactic. Indeed, the tactic in question would not have been employed if the taboo were not sufficiently important to make it worthwhile. If the taboo was irrelevant, then the SFRC's approach would have possessed no political value with which to manipulate the Nixon administration. The example again demonstrates the relevance of fetishization whereby it was the taboo that underpinned the excess value of bioweapons as the basis of the political strategy.

Commitment to the taboo was further reflected in the US's recognition of the Protocol's customary status. Officials accepted that they were 'constrained by a customary international regime' (Moore 1972: 478). The non-use of bioweapons was maintained per the terms of that document. Compliance was again evident within American support for the 1966 resolution in that the US backing of the final text (if not the initial Hungarian jibes) was understood as a formalized acceptance of the Protocol's customary ruling (Synge 1969: 259–60). A SIPRI (1973: 123) analysis emphasized that the resolution

[6] The same argument can be applied to dispute the claims that the US only desired the BWC as a means to restrain others following its unilateral decision to renounce. If this were the case, why would SFRC officials put this 'security' at stake by being willing to sacrifice the BWC over the Protocol dispute, where this would go against the basic idea of self-protection and advantage?

demonstrated a long-standing commitment to the Protocol's customary status and the principles of stigmatization that the agreement expressed. The report stated that it was not the case that the custom started with the resolution in 1966. The obligation in question was already meaningfully entrenched and to think otherwise would be to 'miss the point, for the imperative character of the customary rule which the resolution enunciates does not result from the resolution but from that rule itself' [emphasis in original].

A CIA (1969a) report three years later similarly discussed the Protocol as a customary norm. Furthermore, the CIA analysis recognized that the custom was not limited to the Protocol but also related to a variety of other policy commitments opposing biological and chemical agents:

> Whether all these circumstances add up to a general customary international law banning CBW [chemical and biological weapon] use may be academic. But the hypothesis has exerted pressure on governments over the years, and has molded world public opinion, which now operates as a reasonably effective de facto deterrent against the use of CBW.
>
> **CIA (1969a)**

The Protocol was also explicitly understood in taboo terms. Former Senator and SFRC member, Joseph Clark (in USCong 1971: 195), commented in Congressional hearings that decision-making on this issue was most concerned 'with the moral aspects of this protocol' and on the ethics of the failure to ratify since 1925.[7] Clark said of this failure: 'I don't know where our moral sense has been for those 46 years.'

Taboo advocates continued to influence debate within the Senate and directly through the SFRC. McCarthy (in USCong 1969b: 9) spoke out in favour of the Protocol's ratification as the type of act that would be carried out by 'civilized' persons. Fulbright (in USCong 1971: 1) emotively referred to the disgust of the taboo when he spoke of 'the general abhorrence of chemical and biological warfare' as a core reason to ratify. Advocates also came from within the scientific community and advised the SFRC, including Meselson. Meselson (in USCong 1971: 363) stated in clearly taboo phrasing: 'Public opinion has regarded chemical warfare as particularly uncivilized and repugnant, an attitude that has been shared by not a few political and military leaders.' While Meselson's comments related specifically to chemical weapons, his testimony still placed the taboos concerning poisonous warfare centre stage.

[7] Senator Church (in USCong 1971: 231) also said in the same hearings that any reservation attached to the Protocol would 'see the treaty itself eroded away so that it becomes a much lesser moral force than it is now'. The moral element was, therefore, repeatedly prominent within the decision-making process.

Meselson further highlighted the extent to which public support for the taboo remained an influencing factor. As a Congressional report had expressed the year before:

> Among all forms of warfare, none stands more condemned before the bar of mankind than the use of chemicals and biological agents. From ancient times, employment of poisons has been considered a violation of the rules of war. Today, chemical-biological weapons are looked on by the public with a horror and disgust which is not felt for other, potentially more destructive arms.
>
> **(USCong 1970: 1)**

Ratification, however, was not only an issue of public opinion at the domestic level. The matter also constituted an issue of international sentiment. The HFAC acknowledged that 'chemical and biological weapons are regarded with horror and revulsion by the world' [emphasis added] (USCong 1974b: 370). The committee further stated that this international feeling created a certain expectation on America as a global leader. The US—as a 'major power among the civilised nations'—needed to live up to its 'leadership role in attempts to restrict or ban the use of such weapons through meaningful international agreements' (USCong 1974b: 370).

The debate on renunciation also occurred against the wider backdrop of UN opposition to WMD. The UN commitment to eliminate mass destructive weaponry both strengthened the bioweapons taboo and established an institutional structure for the actual prohibition of these armaments. Removing all WMD had always been the stated objective of this mission. Once the NPT was signed, the BWC was simply a logical next step. The (albeit conditional) prohibition of nuclear weapons created a sense of momentum in which other forms of WMD would now also be subject to elimination. The international community had approved the OST and Seabed Treaties by the time that the US signed the BWC. Both documents explicitly proscribed WMD in specified contexts and so represented the extension of the prohibition norm from nuclear to other forms of mass destructive devices. Consequently, the UN's stigmatization of WMD enabled the diplomatic progression leading to the BWC's conclusion.

The factors discussed above demonstrate that the taboo was a foremost factor in the push towards the elimination of bioweapons. The normative belief that bioweapons were disgusting, unforgiveable, and intolerable armaments drove political thinking and diplomatic action. The realization of the taboo was admittedly conditional. In particular, Vietnam repeatedly muddied the normative waters. Yet the expectation of the taboo was still upheld as a core principle of international politics that must be respected—and where that respect was associated with formal codification in the form of the BWC. Of all the factors that had influenced the decision to renounce bioweapons, therefore, American diplomat, L. Craig Johnstone (1971: 713), concluded that what ranked 'most significantly' were the 'ethical and moral considerations'.

On the books

Indeed, the very expression of the bioweapons taboo within the BWC comprised a key motivation for the agreement itself. Leonard (2001: 14) described the Convention as a case whereby the US could 'get a norm [opposing biowarfare] on the books'. In doing so, Jeanne Guillemin (2005:13) portrays the BWC as a mechanism whereby 'biological weapons were stripped of their legitimacy' (see also McCauley and Payne 2010: 14). The BWC was not merely a prohibition in the minimal sense that it removed certain weapons from international stockpiles. The agreement was a normative endorsement of the taboo. Consequently, the BWC comprised a key juncture within the taboo's genealogy. The previous chapters demonstrate that the taboo had always exerted significant and autonomous sway over political behaviour. Actor rejection of biowarfare was long governed by, and openly acknowledged as, a taboo with respect to decision-making and the appropriate waging of conflict. Even taking this extensive history into account, however, the BWC still constituted a crucial moment when the taboo was very publicly recognized via the formal means of codification (e.g. Goldblat 1997: 264, Millet 2003: 303; Cross and Klotz 2020: 185).

Policymakers have acknowledged the perception of the BWC as a defining marker with respect to biowarfare. For example, the head of the US delegation to the 1986 BWC Review Conference (RevCon), Ambassador Donald Lowitz (1986: 40), specifically described 'the international norm against biological and toxin weapons *established by this convention*' [emphasis added].[8] Of course, the BWC was not the only relevant event in prohibiting biowarfare, as the US renunciation of bioweapons and the Geneva Protocol clearly demonstrate. As indicated by Lowitz's statement, however, the BWC was interpreted as core to the international manifestation of the taboo. The

[8] Lowitz (1986: 40) explicitly described biological arms in taboo terms as 'morally repugnant weapons'. He said: 'And we [the US] continue to be convinced that, because the use of biological and toxin weapons is repugnant to the conscience of mankind, no efforts should be spared to minimize this risk, with the objectives of excluding completely the possibility of such use.'

BWC's statement that bioweapons were intolerable, immoral, and inherently repugnant cemented the taboo to a more profound degree than any other example of normative internalization analysed within this study. Meselson (1999: 56) summarized the situation best when he said of the BWC that 'the international norm [against biowarfare] is unambiguously established'. Significantly, the BWC also went well beyond the terms laid out within the Protocol in that the agreement was now free of all reservations, for example retaliation in kind. The BWC instituted a total ban fully in line with the highest expectations of the taboo ('Telegram 72644', 1971).

The BWC also stands out as a key genealogical moment concerning the taboo in that it established a specific regime prohibiting biowarfare. Elements of that regime were already present, especially in respect of the Protocol and the customary force associated with that document. Yet the BWC created an even more extensive institutional structure and transformed this relatively embryonic regime into a substantially more extensive and influential construct. The regime was now further buttressed by new bureaucratic provisions. For example, the BWC set up regular review conferences that would be rigorously observed (even despite a delay to the 2020 conference due to Covid-19). The new provisions facilitated compliance and regime success in the sense that these rules both (1) established the means by which to more effectively enact the Convention and (2) promoted greater institutionalization of the taboo to further engrain normative internalization. The mutually reinforcing combination of normative commitment and bureaucratic stigmatization still underpins the BWC as a key channel for advancing and evolving the taboo. As such, the BWC is acknowledged as 'the cornerstone of the BW [biological weapons] nonproliferation regime' (Koblentz 2009: 63).

The limitations of the BWC

The BWC is imperfect. Smithson (2004: 165) comments: 'Banning biological weaponry is both prudent and laudable. Putting that idea into practice, however, has been a nightmare.' Elsewhere, the BWC has been called 'fundamentally flawed', 'a bruised paper tiger', and even 'comatose and barely alive' (Beard 2007: 271; Vestergaard and Roul 2011: 489; 'Biological Weapons Convention', 2002: 13). Other analysts interpret the BWC in more complex terms. Littlewood (2007b: 14) describes the criticism of the BWC as a pendulum whereby 'perceptions of the usefulness of the convention have oscillated between seeing it as a beacon of multilateral arms control and viewing it as irrelevant'. Yet the BWC is still 'often considered to be a comparatively weak regime' (Millet 2010: 25). Reproach of the BWC typically centres around two key issues: compliance (whether state parties follow the rules) and verification (the means to ensure that the rules are followed, e.g. inspections).

Compliance

The BWC is criticized on the basis that the Convention has not prevented state actors from developing a biowarfare capacity. *Biopreparat* is frequently cited as a key example of the Convention's apparent failure. The Soviet's extensive bioweapons programme was pursued in clear violation of the BWC. It is even argued that the programme's expansion was facilitated by the BWC's 'lack of teeth' in enforcing the prohibition on bioweapons (Tucker 2010: 9). Moreover—as the reference to Mangold and Goldberg's research already highlights above—some analysts even contend that the BWC actually created an incentive to develop *Biopreparat* once the US and others had renounced (Allen 2007: 5). It was not simply the case that the BWC was unable to preclude violation; the agreement actually encouraged transgressive behaviour.

While the USSR developed their biowarfare capacity in the upmost secrecy,[9] American officials had long speculated that a programme existed. US suspicion was fuelled following a major bioresearch accident at Sverdlovsk (now Yakaterinburg) on 2 April 1979 (Earle 1980: 239). *Bacillius anthracis* was inadvertently released from a military facility based on the outskirts of the Russian city ('Summary of Conclusions', 1980: 243; Guillemin 1999). The incident killed approximately 66 people. Sverdlovsk is described as 'the biological Chernobyl' (Alibek 1999: 80). The Soviets denied at the time that the outbreak was connected to a military programme and instead claimed the infection was caused by naturally contaminated meat. The excuse proved compelling. Even US actors were initially uncertain whether the anthrax was a consequence of military development. While Meselson would later demonstrate that Sverdlovsk was related to a dedicated bioweapon programme (Meselson et al. 1994), he first advised the US government that the Soviets' claims to natural causes 'seemed to hang together' (Meselson in DoS 1986). Meselson was right the second time. President of the Russian Federation, Boris Yeltsin, finally admitted the programme's existence in 1992 (Leitenberg 1997: 8).

[9] Rodney Nichols (2002: 1464) drily applies a 1980 quote on arms control by RAND Corporation analyst, Amrom Katz, to the issue of the Soviet bioweapons programme. Katz had said: 'We have never found anything that the Soviets have successfully hidden.'

Sverdlovsk has been portrayed as a case wherein 'the integrity of the BWC was seriously weakened' (Kearney 1992: 51). The evidence of the USSR's violation ostensibly exemplified the Convention's inherent limitations by revealing a systemic inability of the agreement to police compliance (Gordin 1997; see also Monath and Gordon 1998: 1423; Tucker 2000: 35). The BWC's institutional inadequacy was manifest both as the failure to preclude an initial transgression and as a further powerlessness to bring the USSR to account once the contravention had occurred. It was argued that the inability of the BWC to impose significant consequences for acts of violation meant that the Soviets' had been able to behave with 'impunity' (Rimmington 2000: 2) and that the USSR had manipulated the BWC's provisions in order to avoid sanction (Moodie 2001; Shoham and Wolfson 2004). The US did employ the BWC as a means to challenge the USSR's supposed actions in respect of Sverdlovsk (E. Harris 1987: 46–9). For example, American representatives raised the issue of the anthrax outbreak at the Convention's 1986 RevCon.[10] Yet the claim that Sverdlovsk was indicative of a major programme would not be confirmed until the USSR itself actually confessed. Significantly, the confession came after the Soviets' own decision to make it, as opposed to any direct consequence of BWC measures. The BWC was not the reason that the USSR came clean.

Sverdlovsk was connected to further allegations in the early 1980s concerning Soviet toxin use[11] in Laos, Cambodia, and Afghanistan (Tucker 2001b). One of the alleged toxins used was infamously christened 'Yellow Rain'. The name refers to an amber liquid found at the scene of the supposed attacks against CIA-backed insurgency forces based within the targeted countries. Secretary of State, Alexander Haig (1981), delivered a public statement to the Berlin Press Association on 13 September 1981, outlining the allegations. He stated that US investigations had uncovered 'physical evidence' in Southeast Asia of aggressive toxin employment by the USSR. The accusations were further discussed in a 1982 report by Haig's replacement as Secretary, George Pratt Shultz: Special Report No. 104, Chemical Warfare in Southeast Asia and Afghanistan: An Update (DoS 1982).[12] Yet the validity of the evidence was called into question. Scientific experts, including Meselson, maintained

[10] US Representative at the UN in Geneva, Ambassador Gerald Helman (1980: 259), commented that other delegations attending the RevCon were evidently 'bemused' by the US bringing Sverdlovsk up at such a time and in such a manner, i.e. seven years late and via means of a review conference, which was hardly a world court. Helman additionally noted, however, that 'these same participants also recognized the seriousness of the inquiry and the importance of obtaining a satisfactory response from the Soviets'.

[11] T-2 mycotoxins; a fungi-produced toxin.

[12] The USSR (1983: 79) publicly denounced the report at the UN.

that Yellow Rain was a natural phenomenon; explicitly, the amber substance had been created by the mass defecation of a swarm of indigenous honeybees (Seeley at al. 1985; Robinson, Guillemin, and Meselson 1987; Meselson and Robinson 2008). Genuine or not, US officials repeatedly cited these 'attacks' as possible violations of both the BWC and Protocol (Haig 1981: 46; see also Stoessel 1981: 79, 1982: 57; Burt 1982: 54; Eagleburger 1982: 37; Reagan 1982: 41; Rostow 1982: 44).

The Yellow Rain scandal was again interpreted negatively in terms of the BWC. The alleged violation was said to have 'undermined respect' for, and 'caused some to question the value of', the agreement (E. Smith 1984: 1012; Cole 1997: 179). The BWC had seemingly failed on numerous bureaucratic points pertaining to the alleged Soviet transgression, including an absence of pre-existing multilateral methods for investigation. An official UN inspection into the allegations was encumbered by the USSR's refusal to comply with it (Cassell 1983: 274–7). The lack of a decisive investigation on the part of the UN then engendered a difficult reliance on the US's controversial unilateral analysis for evidence (Tucker 2001b: 38). Moreover, the very fact that the accusations had been made also reduced faith in the Convention. Some analysts even held that America should withdraw from the BWC because the supposed violation had demonstrated that the agreement was little more than a 'sham' treaty (Douglass and Livingstone 1987: 177). Opposition to the BWC was exacerbated even further by additional accusations that the US was only concerned with—indeed, may have actively fabricated—the allegations to facilitate American self-interest. US military officials had been lobbying for an enhanced chemical weapons armoury at this time. Exaggerating an apparent Soviet threat in the biological and chemical field would ostensibly bolster their assertions that America required a credible 'retaliatory capacity' to respond in kind (Weinberger 1983: 61). The claim that the US was only pushing the allegations for their own benefit fortified the criticisms against the BWC.

The USSR is not the sole example of actual or alleged non-compliance: Iraq, Syria, North Korea, Egypt, and Israel (amongst others) all stand accused or proven of bioweapons development since the BWC came into force (Carus 2017a: 28–39). The US has also not escaped accusation. The USSR claimed in the 1980s that HIV/AIDS was an American bioweapon (Shultz 1986a; 1986b). Sixteen years after the BWC was initially signed, Director of Central Intelligence, William Webster (1988), said at the World Affairs Council on this issue that 'it appears that the moral barrier to biological warfare has been lifted'.

Verification

The BWC is further criticized over verification. The concern is now so pervasive within debates on the Convention that it is referred to as the 'v-word' (Findlay 2006: 12). Verification is intrinsically problematic with respect to biowarfare research and development whereby: it is exceptionally difficult to differentiate between permitted and prohibited activities; bioresearch activities are inherently dual use (i.e. the 'same' research can be exploited for both aggressive and benign purposes, the latter including medical advances); and there typically exist few visual signs associated with a programme in that the relevant research frequently occurs indoors, within facilities that are ostensibly permissible under BWC terms, and where it is relatively easy to dismantle evidence of bioscience research if there is any suspicion that inspectors are about to come knocking at your door (Tucker 1995: 11; Smithson 2004: 79).

The UK raised the idea that the BWC was simply unverifiable in their initial draft of the agreement ('British Working Paper', 1968). The British delegation acknowledged from the start that they believed full verification to be a political pipe dream. Subsequently, state parties took the pragmatic decision that an agreement without clear means of verification was preferable to no agreement. The UK draft working paper concluded that 'the choice lies between going ahead with the formulation of new obligations and doing nothing at all' ('British Working Paper', 1968: 570). US officials adopted the same line. Guhin (1971: 1) sent a telegram to Kissinger stating his concern that the proposed agreement would force a reliance on a complaints procedure as opposed to verification, that is, the agreement would effectively deal with any incidence of non-compliance after the fact instead of ensuring that it would not happen. He added, however, that the prospects of securing an international consensus on the BWC quickly were too 'good' to ignore. It was best to act now, even if the document was unsatisfactory on the issue of verification. Leonard (1986: 12) said: 'The Nixon administration believed that a Soviet commitment not to possess biological weapons even without verification, would be better for U.S. national security than leaving the Soviet Union and others legally free to experiment.'

Repeated efforts to resolve the verification question have ended in failure and diplomatic disharmony. An expert committee entitled VEREX was established at the 1991 Third RevCon to examine the potential introduction of new verification provisions. The group's subsequent report was debated at a Special Conference in 1994. The VEREX recommendations led to the setting up of a further committee of governmental experts, the Ad Hoc Group (AHG), to negotiate a draft protocol on verification. The AHG discussions quickly descended into acrimony. Yet the AHG Chair, Tibor Toth, was finally able to put forward a compromise text for approval in 2001 (Kelle 2007: 221–2). The document proposed a significant increase in verification measures. The suggested measures, however, would still be subject to considerable restrictions including the option by state parties to challenge investigations, time limits on inspections, and constraints on the types of scientific equipment and technology inspectors could request to analyse.

America said no to Toth's proposal. Furthermore, the US did not simply reject the verification protocol. America additionally moved to terminate the AHG and effectively shut down this specific pathway to new provisions.[13] US delegate Donald Mahley (in 'U.S. Rejection', 2001) explained that the modes of verification which had been proposed were simply unworkable. Hence there was no point pursuing them further:

> In short, after extensive analysis, we [the US] were forced to conclude that the mechanisms envisioned in the Protocol would not achieve their objectives, that no modification of them would allow them to achieve their objectives, and that trying to do more would simply raise the risk to legitimate United States activities ... we are forced to conclude that an appropriate balance cannot be struck that would make the draft Protocol defensible as an instrument whose utility outweighs its risk.
>
> **(Mahley in 'U.S. Rejection', 2001: 901)**

There was concern within the American delegation that the proposed verification protocol would permit other states to harass US efforts on vaccines and defensive measures, threaten scientific and commercial secrecy, and would also prevent the US from enforcing regulation controlling the export of biological-related technology. Officials were also under specific pressure from American pharmaceutical companies who considered greater verification restrictions to be unnecessary, inconvenient, and financially incautious (Leitenberg 2002: 4; Guillemin 2005: 168).

How far the protocol's failure was the fault of the US remains subject to dispute. John Borrie (2006: 18–19) specifically attributes the situation to a lack of US leadership. Alternatively, however, American involvement in the protocol's demise has been described as 'a convenient scape-goat for many to protest the perceived "unilateralism" of the Bush administration or to conceal their own culpability for the impasse in the negotiations' (Ward 2004: 198). Verification is portrayed here as being inevitably lost in a compromise

[13] For a full assessment of the verification protocol's collapse see Littlewood (2005).

text that was so controversial it would never have gained the necessary consensus to pass regardless of American input. The US was merely hammering a nail into a coffin lid that was already very securely pinned down. Either way, the verification issue remains unresolved. There still exist 'persisting controversies' concerning the supposedly 'toothless' agreement on this point (Tucker 1997; Drobysz 2021). Moreover, it is argued that the identified problems concerning verification are so extensive that they will ultimately result in the downfall of the entire Convention (Steinbruner 1997: 89). The inability to verify will not only permit the actual proliferation of bioweapons but will also undermine international trust and multilateral diplomacy. While poor verification does not necessarily equate to poor compliance, it is still crucial for other aspects of international political interaction including the deterrence of violation, creating a multilateral platform for future diplomacy on biowarfare, and building confidence between state parties (Chevrier 1990: 95). Failing to achieve trust within the BWC regime then creates a risk that the Convention will become 'irrelevant' (Kervers 2003: 161).

Limits on the limitations

The BWC's limitations have been connected to the taboo (McCauley and Payne 2010: 6–7). To quote Smithson (2004: 173) again: 'For the time being, the international community appears to remain a culpably negligent caretaker of the norm against biological weapons.' Malcolm Dando (1997: 2) refers to the bioweapons' prohibition norm as 'by far the weakest' of all those pertaining to forms of WMD. The identified controversies concerning the Convention 'could undermine the taboo against bioweapons, giving the impression that the BWC is ineffective' or is 'limited in its effectiveness' (Bailey 1995: 384; Meier 2002: 183). Similarly, James Revill (2017: 7) argues that norms are manifest as social constructs which are vulnerable to change. Within this context, Revill further claims that the BWC is not designed in ways that are sufficient to resist and preclude the pressures of change that threaten to collapse both the regime and the taboo itself.

Yet Revill (2017: 12) also recognizes that the taboo is currently 'widely recognized and seldom contested'. Other studies support a positive interpretation of the taboo's robustness. Michael Horowitz and Neil Narang (2014: 529–30) demonstrate that BWC membership is associated with lower actor interest in bioweapons development. The authors specifically attribute this behaviour to a prohibitive normative belief which opposes biowarfare, manifest within the BWC. Furthermore, Dando's claim that the BWC regime is relatively weak is also disputed. Jason Enia and Jeffrey Fields (2014: 44, 52–3) claim that the BWC regime is in fact superior to the one relating to nuclear armaments. They state that the Convention has achieved more than the NPT in delegitimizing the weapons which the respective agreements seek to preclude. Is the taboo, therefore, really at stake in respect of the BWC's deficiencies?

Concerns of non-compliance must be contextualized. Violations have occurred but they are 'few and far between' (Cross and Klotz 2020: 189). Arthur Westing (1989: 82) pragmatically comments that 'transgressions of these culturally imposed and legally codified prohibitions are bound to occur from time to time in the future'. He further adds that even limited deviance has not challenged the normative concern surrounding bioweapons. The rejection of bioweapons remains 'deeply ingrained in the human ethos' (Westing 1989: 82). John Parachini (2001) similarly asserts that arms control prohibitions 'have, of course, been violated at times by certain states that had pledged to uphold them, but they still matter'. Parachini makes his comments explicitly within the context of the BWC and the idea that the Convention normatively prohibits a pariah weapon. Consequently, concerns surrounding the BWC's terms do not automatically entail that certain ideas (even those we might have hoped to be ideal) are invalid to any decisive degree.

Transgression may also only be perceived. Allegation does not mean that a violation has actually occurred. The BWC was heavily criticized following the Yellow Rain claims. Yet the subsequent condemnation of the agreement was largely dependent upon an assumption that the allegations were true—which is a questionable understanding at best. Critics were responding to a sense of transgression that cannot be decisively shown to have happened, that is, that the BWC and the taboo did not necessarily fail in this situation. The statement here is not to negate the actor's response but to argue that, if the act did not happen, the criticism that both the Convention and the taboo did not prevent that act cannot be logically applied. Furthermore, contravention (whether alleged or real) has not generated any wider domino effect. For example, the US did not re-build its bioweapon programme even after receiving confirmation of *Biopreparat*. Violation does not automatically signal that the terms of the BWC and the taboo which it expresses will not hold.

Violation has even promoted the taboo, as claimed in Chapter 2. Moreover, the BWC has provided the political mechanisms for doing so. Sverdlovsk has been portrayed as a failure of the BWC's procedure. This argument, however, overlooks a critical factor in that this procedure *was* employed. US actors deliberately selected to pursue their complaint via the BWC regime as opposed to by purely unilateral or alternative means ('Editorial Note'

1980; Vance 1980: 770). Sverdlovsk was also specifically constructed as a normative violation of the BWC. Officials even asserted that the incident must be denounced as such specifically in order to ensure that the BWC was not undercut. Acting Secretary of State, Warren Christopher (in 'Memorandum of Conversation', 1980: 852), said of the Sverdlovsk issue that he explicitly wished to 'not permit this question to undermine the BW convention' (see also Christopher 1980: 281). Consequently, Sverdlovsk: emphasized the international community's commitment to the taboo; reinforced a global recognition of the BWC as the institutional expression of taboo and the primary political vehicle for sanction pertaining to normative violation; and directly addressed Sverdlovsk as an act of violation (in contrast to the claims elsewhere in the literature that transgression in respect of biowarfare has effectively been ignored).

Normative reinforcement also relates to the sense of violation associated with the Yellow Rain allegations. ACDA Director, Eugene Rostow (1981: 70), said that Yellow Rain 'makes clear the need for concrete international action to restore confidence in' both the Protocol and the BWC. Rostow's comments further demonstrate the dynamic of non-compliance in strengthening the taboo whereby he called for a strong response to the perceived contravention. His desired response explicitly sought to both demonize any deviation from the normative standard and demand an even more robust stigmatization in respect to future acts of biowarfare. This commitment to the taboo was not exclusive to Rostow. The US reaction was repeatedly phrased in terms of a need to respect the BWC and the normative prohibition against biowarfare that the agreement institutionalized. Under Secretary of State for Political Affairs, Lawrence Eagleburger (1983: 77), said that the need to show respect for the Convention was particularly pertinent to the US given that America had 'played a major role in negotiating an international agreement banning these weapons' and so was required to take the matter seriously. Eagleburger further constructed his argument within taboo terms when he stated that the BWC comprised the 'universally accepted standards of international law and respect for humanity'. President Ronald Reagan also adopted a taboo stance. The president discussed the Yellow Rain allegations in terms of expectations of civilized and humane conduct. He disparaged the Soviets' actions in using these 'nightmarish' weapons as being carried out 'with impunity', 'brutality', and even compared the situation to the Holocaust: 'I don't think that we can remain silent, as too many of us in the world did when Hitler was coming to power, in the face of this kind of conduct' (Reagan 1983a: 24, 1983b: 28, 1983c: 66, 1983d: 447). Far from undermining taboo, therefore, the US response to a perceived violation reinforced the need to defend the norm.

The same dynamic of normative reinforcement is evident in the domestic US outcry concerning internal failures to comply with BWC-related measures to eliminate bioweapons and weaponizable agents. It was revealed in 1975 that the CIA had retained limited stockpiles of biological agent including shellfish toxin and cobra venom. The CIA's reason for keeping the agent is unclear. Officials put forward a range of contradictory explanations: that they 'could not bear' to destroy their research (Winchester 1975: 2); that the failure to destroy the agent was due to the excessive cost of doing so (in USCong 1975: 7); and that the CIA had simply not received instructions to remove the agents, despite a clear presidential order to do so (USCong 1977: 86). Whichever one (if any) of these excuses is correct, the retention proved highly controversial and provoked a substantial backlash. The negative response included a vehement condemnation of the CIA's actions during Congressional hearings (USCong 1975, 1977). Similar acrimony was expressed in response to disclosures that the US Army had withheld eight deadly toxins in violation of Nixon's directive to destroy them ('Storage of Toxins', 1975: 16).

Further outrage ensued following a media exposé that the CIA had carried out secret bioweapon testing in 1966 on the New York subway. Scientists had filled light bulbs with harmless *Bacillus globigii* bacteria and smashed them open on the train tracks in order to analyse how the agent travelled through the transport system ('Subway', 1975: 27). The experiments themselves took place prior to renunciation. Yet the disturbing and lurid tales of clandestine public testing now played into wider concerns that reflected the taboo in an environment of committed bioweapons elimination. Consequently, these examples again demonstrate that the idea of violation did inspire condemnation, sanction, and (at least a commitment to) increased measures upholding the normative rejection of biowarfare, explicitly where this taboo sentiment was related to the BWC.

Moreover, the negative focus on verification has been said to ignore the influence of the taboo in terms of the BWC. Freeman Dyson (1984) commented in his book, *Weapons and Hope*:

> The chief purpose of the convention is to strengthen the traditional consensus of mankind that biological warfare is in some sense peculiarly evil and dishonorable. The value of the convention for this purpose remains largely intact, whether or not it is verified and whether or not it is violated.
>
> **(Dyson 1984: 176)**

Dyson identifies the taboo as the primary function of the BWC. Specifically, Dyson understands the commitment to the taboo as integral to assessing whether the agreement is considered a success. He argues that the bureaucratic details of that function's enactment (such as verification) should not detract from the Convention's prioritization of the taboo. The taboo and the principles that the BWC expresses are important and make a contribution to actor behaviour, even if the precise provisions of the agreement are lacking. Dyson (1984: 177) said: 'The biological weapons convention illustrates the general thesis that arms control agreements do not have to be perfect to be useful.'[14] Furthermore, to the extent that the BWC has been dismissed as too simplistic and lacking in more detailed provisions such as verification, Nicholas Sims (2007: 352) has stated that the agreement expresses a 'simplicity and comprehensiveness that, properly understood, are sources of strength not weakness'. An uncomplicated agreement can be more effective in upholding a taboo.

It is additionally shown that the BWC cannot be viewed in isolation from wider aspects of the anti-biowarfare regime and WMD non-proliferation generally. The BWC 'no longer stands alone as the sole bulwark against biological weapons' in that the Convention is now supported and complemented by other facets of non-proliferation including the CWC, the Group of Eight, Global Partnership, and the UN Secretary General's investigation mechanism pertaining to biowarfare development (Sims and Littlewood 2011: 501)—in the vein of a norm cluster. Consequently, the problems that have been identified pertaining to verification are mitigated by wider normative support. Indeed, the relative absence of biowarfare in spite of poor verification is a strong indicator of norm robustness. Actors have adhered to the taboo in not using bioweapons even though they can get away with it because verification is weak. The barn door is open, but no horse is bolting through it. Far from undercutting the normative standard opposing biowarfare, therefore, the deficiency pertaining to verification has actually been kept in check by the very strength of the taboo.

Becoming robust

The BWC is clearly flawed. The Convention, however, still has its defenders. It is maintained that the BWC still exerts substantial political weight in that the agreement 'remains strongly supported by the international community'

[14] Paul Meyer (2018: 31) similarly says of WMD treaties generally that prohibition 'remains imperfect, but nonetheless represents a major global public good, and an inspiration for those who believe in the potential of cooperative security arrangements'. He adds specifically that these treaties do have an 'influence on state conduct'.

(Graham 2002: 1236–7) and also provides a critical 'normative centre' to the biowarfare regime (Fidler and Gostin 2007: 48; see also Cross and Klotz 2020: 185). Specifically, it is also not clear that the deficiencies identified in respect of the BWC have decisively undermined the bioweapons taboo. While Littlewood (2007b: 14) bemoans the problems of verification, he also does not deny that the 'norm against state use of biological weapons and the legal embodiment of that norm, the BWC, are strong'. Leonard (1986: 13) similarly highlighted the success of the Convention on these same grounds: 'No government has denounced the norm against the use of biological weapons or withdrawn from the treaty. ... Virtually all governments remain convinced that the BWC must be protected.' Zanders (2001: 365) takes this claim even further to maintain that the universality of the agreement's expressed norm constitutes a great strength of the Convention. Following Dyson, Zanders argues that this identified strength 'partially compensates for the reduced potency of the negative security guarantees as a consequence of the limited verifiability and enforceability of the convention'. Consequently, the BWC still comprises a fundamentally robust expression of the taboo.

Overall, this chapter is not intended as a defence of the BWC or a decisive rebuttal of the criticisms made against it. The analysis does, however, uphold a defence of the presence and relevance of the bioweapons taboo. Explicitly, it is argued here that the BWC has proven to be integral in upholding and institutionalizing the taboo as an act of codified stigmatization. The BWC also contributes to fetishization surrounding bioweapons. By highlighting biowarfare as a taboo concern of the highest international interest through the BWC, so the disgusting nature of these weapons—manifest within the need to preclude them at all costs—has intensified the extreme 'value' associated with these weapons, which underpins the taboo.

Critically, this study (once again) does not claim that the bioweapons taboo is now fully robust. The BWC is distinguished here as a core marker in the taboo's genealogical development; yet the identified importance of the Convention is not equated with the statement that the agreement produced an ideal taboo which has never been questioned or has not evolved since 1972. Everything that has happened since that date, however, has been shaped not only by the persistent repulsion that we feel towards bioweapons but also by the way in which the world came together to formalize that sentiment within the codified document of the BWC. In this landmark agreement the international community chose to actively declare that biowarfare was illegitimate. And these actors did so in the name and the words of the taboo. The terms of the taboo's political recognition and our understanding of the norm may, of course, change in the future. The BWC is not the end of the story. In terms of biowarfare and the political aversion to bioweapons today, however, in so many ways the story starts here.

Conclusion

Covid-19 has demonstrated the destructive power of disease. The pandemic exemplifies not only the physical effects of infection and death, but also the myriad ways that sickness can disrupt both national and international societies. Lockdowns, closed borders, and restrictions upon our rights and freedoms—are all examples of the means by which disease can undermine the very way that we live. In showcasing what disease can do, the pandemic additionally highlights the threat of biowarfare and what could occur if disease were to be strategically utilized as an offensive weapon. As former US National Security Advisor, John Bolton (in Budryk 2020), commented: 'Whatever the source of this pandemic, it's a roadmap for the people who do control biological weapons, or aspire to biological weapons, what can happen.' Admittedly, Covid-19 also reveals the potential difficulties associated with bioweapon employment. For example, the outbreak demonstrates how contagious disease can spread and is problematic to contain to a specific target. Yet Covid-19 still comprises a potent reminder that bioweapons represent a substantial threat. The threat is one that we must fully understand if we are to preclude biowarfare from becoming a reality. But we have not understood it. In particular, we have not understood how the emotive and visceral perceptions that we hold about biological armaments play into the construction of security policies and fundamental questions as to whether actors would even employ this mode of aggression. More specifically, we have not understood if and why bioweapons are taboo.

This book has explained and provided an understanding of the bioweapons taboo. In doing so, the study has further demonstrated that the wider analytic comprehension of 'taboo' has been inadequate. The book has challenged current approaches towards the conceptual identification of a taboo as (1) having yet to fully articulate the normative characteristics of a taboo and (2) being too dependent—explicitly or implicitly—on compliance as the primary analytic standard. Consequently, the book has identified a critical and problematic gap in the study of taboos. In response, the book has proposed a more comprehensive and effective model of the concept. The model

The Biological Weapons Taboo. Michelle Bentley, Oxford University Press. © Michelle Bentley (2023).
DOI: 10.1093/oso/9780198892151.003.0009

defines a taboo as a prohibitive norm that: *disgusts*, is *stigmatized*, is *fetishized*. If a norm fulfils these necessary criteria, then it can be considered to be a taboo.

The model applies even in respect of a taboo's violation. Critically, the model does not deny the extent to which compliance remains integral to the conceptualization and measurement of a taboo. A taboo still comprises a norm for which the need to adhere is important—as is the case for any norm. Analysis can apply the proposed criteria of the model and still accept that, as Patricia Greenspan (2015: 118) expresses it: 'A taboo is not something we are *supposed* to apply selectively' [emphasis added]. The model also does not dispute that extensive and repeated transgression can undermine a taboo to a point at which we could consider that normative rule to no longer exist. Furthermore, the model is not considered here to express an essentialist definition of a taboo. The key contention of this study remains, however, that our analytic conceptualization of a taboo must be explicitly and more fully articulated in terms of the entire range of characteristics that define it, of which compliance is only one element and not exclusive.

The book has applied the model to the bioweapons taboo in the first comprehensive analysis of the prohibition. The scope of the taboo's impact has been interpreted genealogically as a conception of evolving normative force. The taboo has long exerted an autonomous normative significance. Yet the book has also identified a core period within the overall development of the taboo's robustness. The selected timeframe opened with the first truly transformative codification of the taboo through the Geneva Protocol. More than any previous normative expression or official document, the stigmatization of the taboo embedded within the Protocol recognized the rejection of bioweapons as a formal statement of prohibition—albeit conditionally at this stage. The taboo was now entrenched as a key feature of international political thinking and codified expectation. The Protocol also provided the clear grounds for any international response should a transgression occur (although, notably, such a transgression never clearly did) and established the taboo as a customary restriction. The Protocol strongly affirmed the taboo as a workable political convention.

The taboo came under pressure during World War II. The US bioweapon programme was instigated in response to the global conflict and also the Cold War that was to follow. Yet the strategies underpinning the American programme were still decisively shaped and constrained by the normative expectations associated with the taboo. Security policies were restricted to defence only as offensive options were deemed morally impermissible. The taboo additionally dominated the rhetoric that was employed around the programme, even by those actors who did not wish to uphold the taboo's principles. Moreover, the biowarfare programme was highly contentious and incited severe public backlash in reaction to its very existence. Consequently, the hypothetical consideration of biowarfare options behind closed doors at a time of extreme national threat was one thing. The wider acceptance, legitimation, and enactment of those ideas within the conflicting context of taboo sentiment was quite another.

Commitment to the taboo was heightened during a further conflict: the Korean War. Communist allegations that the US had employed biowarfare tactics at this time exemplify the taboo's normative significance. The taboo was central to the propaganda. Yet Communist actors would not have utilized the taboo as a basis of their strategic narrative if that norm did not exude sufficient normative power to compel and coerce others. Nor would US actors have similarly employed the taboo in order to retaliate rhetorically against the Communist attack if they did not believe the same. Indeed, the allegations further enhanced the taboo's relevance via the public repetition of its ideals within this identified propaganda and the strategic exploitation of the norm's associated disgust. Lurid tales of infectious insects exacerbated the repulsion that is intrinsic to the taboo. Admittedly, other factors would continue to effectively clash with the taboo. For example, anti-taboo military actors were instrumental in the introduction of a first-use biowarfare policy in 1956. The taboo, however, was still intensifying in terms of its robustness and political sway.

The US officially renounced bioweapons in 1969. Previous analyses of renunciation have attempted to clarify the policy of renunciation primarily in terms of military utility. This book has put forward an original and alternative explanation that renunciation was in fact a consequence of the taboo. Even contemporary references to military utility drew upon the taboo in that the argument was based on a rejection of biowarfare as causing indiscriminate and mass destruction, which are both core taboo themes. Moreover, claims pertaining to utility were often simply a mask to convince a hard-line military audience of the pro-taboo position. The taboo was further evident regarding: the extreme public pressure resulting from the controversies of Vietnam and other scandals; the persuasive impact of taboo advocates who both facilitated and exacerbated public and political opinion; and international expectations, such as the stigmatization of WMD at the UN.

The same normative dynamics would ultimately push the US towards a multilateral commitment to proscribe biowarfare in the form of the BWC. Genealogically, the BWC constitutes the primary marker in respect of taboo robustness. The Convention is far from perfect as an international pact or as a diplomatic manifestation of the taboo. The agreement, however, still represents the most important and compelling statement relating to that normative prohibition. The BWC comprised the point at which the complete rejection of biowarfare (in contrast to the reservations of the Protocol) became engrained as a staple of international politics; and, more explicitly, as an expression of the taboo. The commitment to the taboo evident within the BWC still reverberates in terms of biowarfare politics today and constitutes a central reason why bioweapons remain a fundamentally hypothetical concept: something to be feared, but never enacted.

Disgust, stigmatization, and fetishization

In applying the new model of a taboo to this history, the book has shown that disgust comprises a decisive factor in relation to biowarfare and actors' rejection of biological tactics. Bioweapons are intrinsically repellent. Revulsion is connected to both the potentiality of biological armaments for mass destructive effect and a fear of disease as an abysmal, invisible, and indiscriminate contaminant. As the reference to indiscriminate force in particular attests, disgust-driven aversion towards contamination does not exist as a purely biological consideration, but also as a social one. Actors could not sanction any resort to biowarfare because sentiments of revulsion prohibited such action as immoral and uncivilized. Disgust is evident in (amongst so many examples within this book) the taboo talk of: Leahy's expression that biowarfare was so repulsive as to 'violate every Christian ethic'; Stimson's 'dirty business'; Gross' claim that the world was 'sickened' by the biological means of aggression; McCarthy's statement that bioviolence is 'directly contrary to the principles of humanity and decency for which America stands'; and even the BWC's own official declaration that bioweapons are 'repugnant to the conscience of mankind'.

The book has also demonstrated the significance of stigmatization to the bioweapons taboo. Stigma negatively differentiates the perceived abnormal as the basis for rejection. Biowarfare has been consistently stigmatized in this vein as the emotive and deliberate inverse of the normal and uncontaminated body. Biowarfare is further segregated on the grounds of its constructed deviance from modes of conventional aggression. The same features that make bioviolence disgusting (invisibility, invasiveness, infectiousness, and indiscriminate impact) are similarly the justifications upon which these weapons are differentiated from the conventional and so stigmatized as taboo.

Stigmatization is especially evident within the influential WMD norm developed at the UN since 1945. The book has outlined how the UN stigmatized bioweapons as part of the wider WMD categorization within the institution's very ethos and mission. Crucially, the UN's normative expression of WMD was not merely a statement of taboo feeling. The taboo was incorporated within the institutional structure and bureaucratic fabric of the UN itself as a foundational principle of the post-war international community. In line with Tannenwald's claim that world opinion constitutes a key source of normative proscription, the taboo was entrenched as a priority concern at the deepest levels of global politics. Furthermore, the collective nature of the WMD norm engendered a sense of momentum pertaining to weapons prohibitions that favoured the bioweapons taboo. Once the NPT was signed, attention automatically turned to the elimination of other forms of WMD, including bioweapons. Efforts to negate this taboo with reference to biowarfare—for example, the Stevenson Committee's questioning of the WMD concept's relevance to bioweapons—repeatedly failed. The UN stigmatization of the taboo comprised a core aspect of biowarfare politics that US actors could not snub, even if they so wished.

Stigmatization is additionally manifest within the taboo's codification in respect of international agreements. Codification is not considered here to constitute a necessary aspect of a taboo. The book has shown, however, that the formalization of normative restraint within official proscriptive agreements has significantly clarified and reinforced the bioweapons taboo. Codification gave power to the taboo and established it unequivocally as a central standard of both international and domestic political thinking. The book has identified the two key markers of the taboo within its genealogical development: the Geneva Protocol and the BWC. It is no coincidence that these two markers are also major examples of international codification. Admittedly, the extent and efficacy of both the Protocol and the BWC can be questioned. Yet both documents represent a significant step forward in terms of the taboo's political recognition and robustness. As has been repeatedly asserted in the book, this statement is not to suggest that the BWC is some endpoint of the taboo. This study finishes on the BWC not as an end but as a landmark move towards a more robust taboo. Yet the codification inherent to that robustness is still key—not as the entirety of the taboo, but as a critical mainstay of its normative influence.

Furthermore, this book has demonstrated that fetishization comprises an essential and intrinsic aspect of a taboo. While fetishization has been drastically overlooked in respect to taboos, the dynamic has been incorporated here as a necessary element of analysis. Three key issues relate here. First, fetishization explains forms of taboo that pertain to the sacred. Sacredness is highly relevant in terms of bioweapons in that disease is heavily associated with apocalyptic and Biblical connotations, per the First Horseman in the introductory chapter. The connection between sickness and the sacred directly contributes to the taboo repulsion experienced towards biowarfare. The taboo talk outlined in this study frequently draws upon related ideas that biowarfare constitutes an immoral and uncivilized act that puts humanity itself at stake and where this expression often has an 'end of days' quality to it. This theme is especially evident in respect of the allegations during the Korean conflict. Biowarfare was constructed here as a clash of 'good versus evil' which reflected back onto the politics of the Cold War as an ideological battle for human civilization and the 'right' way to live. Consequently, perceptions of the sacred cannot be ignored in respect of biowarfare and the taboo.

Second, fetishization expresses the concept of disgust-as-desire. IR's current reliance on stigmatization in conceptualizing taboos means that the discipline cannot address any aspect or functional attribute of a taboo beyond a very basic and inadequate notion of negative rejection. Yet disgust can attract as much as it repels. Such desire may be inherently unwelcome. Like Leontion, we cannot resist staring at the horror in front of us however much we might wish that we could turn away. Desire, however, may also be manifest as a type of lust and paradoxical enjoyment in respect of the sordid. Significantly, desire would not necessarily impel an actor to break a taboo. Normative constraint still holds. Fetishization, however, explains the full gamut of emotion felt towards a taboo subject. Actors find biowarfare abhorrent—but also intriguing, titillating, and exotic. Incorporating fetishization into the analytic examination of the taboo here, therefore, makes a significant contribution to our understanding of this type of norm by demonstrating the entire range of sentiments that underpin decision-making. For example, the allegations during the Korean War were compelling not only because they were horrific but also because they were simply fascinating. Specifically, the interest that the allegations inspired contributed to the recognition of biowarfare as a major international concern and placed the accusations prominently on the political agenda. Repulsion is not merely rejection, therefore, but excitement.

Third, the book demonstrates that the fetishized construction of the taboo attributes an excess value to bioweapons. The distinction of bioweapons is especially manifest within their classification as WMD. The WMD term deliberately seeks to differentiate certain weapons as something 'more' than their conventional counterparts as a type of greater value. Biological armaments are then othered compared to what we understand an ordinary weapon to be. More explicitly, bioweapons are not deemed ordinary as a consequence of their taboo status. The stigmatization inherent to the BWC is not merely a practical or technical matter of outlawing a specific type of violence. It is also not the case that the BWC was negotiated solely in response to the mass destruction associated with bioweapons. Instead, the prohibition comprised a significant statement concerning the extreme value and political weight attached to these armaments. The argument here is not to claim that the dynamics of normative importance and destructive capacity are not interlinked, not least where the latter relates to the critical issue of disgust. The book has shown, however, that biowarfare comprises more than a fear of destruction. The threat also constitutes the extreme distinction of bioviolence as a taboo. In understanding the taboo, therefore, analysis of bioweapons should not be treated merely as an exercise in examining abhorrence, but explicitly an abhorrence *that exceeds others*.

Overall, taboo sentiment was clearly stronger for some actors than for others. Moreover, the rhetorical expression of the taboo was often strategic and employed by actors with additional motivations to do so (e.g. the propaganda of Korea) or who did not personally endorse the taboo (e.g. Operation Blue Skies). Yet even such calculated rhetorical engagement of the taboo still serves to exemplify that norm's presence and influence. As already highlighted in relation to the case study of the Korean Was, the taboo was an attractive basis for propaganda explicitly because the norm was so compelling; and that the allegations themselves reinforced and further instilled the aversion surrounding biowarfare, increasing taboo robustness. Similarly, the US pro-biowarfare lobby's decision to employ the taboo and taboo themes within their official rhetoric of Operation Blue Skies demonstrates normative relevance. Regardless of their own opposition to taboo expectations, military actors (notably the CC) employed rhetoric which acknowledged that: the taboo exuded extensive normative relevance; other actors' acceptance of the taboo could not be ignored, thereby creating a pressure to at least be seen to comply; and the taboo was so powerful that it possessed the capacity to threaten their position to pursue bioweapons. If actors believed that they had to effectively respond to the taboo, it only demonstrates how significant that taboo was.

Pro-biowarfare proponents never succeeded in overcoming the taboo entirely. Actors may have achieved limited victories, such as the establishment of a wartime programme and the introduction of a first-use policy in 1956. Yet neither the programme nor the policy lasted. Related strategy was also highly constrained in respect of the taboo. The repeated adoption of a defence only standpoint rejected offensive use as intolerable in line with the ideational and behavioural demands of the taboo. Furthermore, biowarfare revelations such as the publicizing of the World War II programme through the Merck Report met with a vitriolic public response once these issues left the realm of military secrecy and became subject to wider scrutiny. It should be noted, however, that this pro-taboo opinion was not purely a case of public sentiment. The taboo has been upheld even within the military. As Westing (1985a: 631) comments, a key 'military restraint against biological warfare is that military officers widely consider the use of biological weapons dishonourable, unethical, or treacherous'. Ultimately, moreover, the US was never proven to have used bioweapons. It has not been decisively shown that American biowarfare proponents ever won outright against the taboo in terms of the actual and extensive military employment of biological devices. In fact, these actors often strengthened the taboo by repeating and engaging with the norm and the disgust which the taboo expresses. Taboo talk, therefore, is very much not cheap talk—even when it is intended to be.

Implications beyond the study

The book has investigated US biowarfare policy over a specific historical period. The analysis, however, has generalizable implications that relate beyond this. First, the proposed model of taboo is applicable to all taboos and not solely that pertaining to biowarfare. The model provides a basis for assessing and re-assessing our analytic understanding of what taboos mean more generally and in respect of political decision-making concerning an extremely wide range of issues. These issues include, for example, the prohibitions characterizing other types of weaponry, especially other forms of WMD. Adopting a framework of analysis which is more open to normative transgression and conditionality creates new opportunities to revisit what is already known about other armaments identified as mass destructive. Furthermore, the 1948 UN definition of WMD is not limited to nuclear, biological, chemical, and radiological weapons, but refers to 'all other major weapons adaptable to mass destruction'. Consequently, debate concerns not only the WMD that we recognize now, but the WMD that could be. New weapons may emerge; existing armaments can evolve and become a pariah in our minds. Within that context, this study offers an analytic template for understanding why and how weapons may become taboo in the future, the criteria which would underpin that construction, as well as what it will mean politically and in respect to conflict if this occurs.

The model further informs analysis concerning weapons not currently deemed to be taboo. Weapon taboos are controversial, as is the labelling that accompanies them. The concept of WMD has been criticized whereby the term creates contested hierarchies around how actors consider and respond to conventional versus unconventional violence (Bentley 2014, 2016). This book has deliberately not passed judgement on whether the bioweapons taboo is a 'good' thing as a consequence of this concern. Some analysts do uphold the taboo as a positive bastion against the employment of biological armaments. Paul Schulte (1999: 329) refers to Lederberg as stating that 'there may be no technical solution to the problem of biological weapons, only an ethical solution'. The taboo can be related here to references to the ethical, whereby the taboo would then be seen as critical to non-use within the context of Lederberg's argument. If we cannot control bioweapons from a technical perspective, then the taboo seems to offer our only hope. Lynn Hansen (1990: 54) similarly cautions against the BWC's abandonment on the basis of its multiple flaws (outlined in Chapter 7) in that this rejection could undermine the moral commitment against biowarfare, which he values: '[T]he moment a nation loses sight of the moral imperative, its political strength and vitality begin to wane.' Again, Hansen's reference to morality can be equated with the taboo as a key force ensuring that bioweapons are not deployed. Yet the marking out of select weapons as pariah risks overlooking the horrors of other, conventional, arms. Furthermore, the hierarchy between conventional and WMD armaments created by the definitional structure can even be seen to legitimize conventional weapons, thereby increasing the chances that conventional aggression will occur. The model of the taboo proposed within this book can assist in comprehending how these hierarchies form and function with respect to security beyond biowarfare—and how beneficial or detrimental those constructions may be.

Second, the model can be applied to other countries. This study has recognized, but not fully addressed, several other state biowarfare programmes in addition to the US, for example the World War II Japan and *Biopreparat* regimes. Such examples ostensibly challenge the taboo in that they represent the development of major biological programmes. Yet there is indicative evidence that the taboo on bioweapons remains applicable. As this book has demonstrated, the Japanese Emperor was not informed of the programme for fear of associating him with a contentious form of aggression. US officials also believed that the USSR was still driven by political norms that acted as a constraint. In seeking to understand these cases in greater depth, this book can provide a prototype for unpicking such examples to locate even partial evidence of the taboo. The template, however, applies not only in respect of the programmes that were pursued but also the many programmes around the world that never happened. Is the absence of bioweapons always a presence of the taboo, whereby this can be identified as per the terms of this book? Or is some other factor at play? Critically, this observation is not to claim that the taboo will necessarily hold for all states in precisely the same way. In particular, future analysis should engage with explicitly non-Western approaches which can take into account the limited nature of the US focus adopted here. This study, however, still provides a valuable guide—or at the very least, a point of comparison—for that analysis.

Third, the conclusions of this book are applicable beyond the time frame examined. The study is limited to a historical examination up to *c.*1972. This restriction permits the detail of scrutiny required to fully ascertain the composition and scope of the taboo's influence by analysing core case studies in depth. What this historical analysis reveals, however, is transferable to a more modern context. The taboo is just as disgusting, stigmatized, and fetishized today as it appears in the pages of this study. President Barack Obama's 2009 *National Strategy for Countering Biological Threats* reaffirmed a commitment to the BWC as the 'premiere forum' for preventing bioweapon use and argued that the taboo enshrined within that agreement comprises the key channel to block biowarfare (US 2009: 8–9, 19–23). The administration of President Donald Trump adopted a similar approach with the 2018 *National Biodefense Strategy* (US 2018: 14). Consequently, actors still recognize the taboo as a core means for the successful preclusion of biowarfare. Continued commitment to the taboo can be investigated within the terms laid out in this analysis, where the study can explain the following: the ways in which the taboo restrains and shapes actor behaviour; the intrinsic character of the taboo in doing so; the extent to which the taboo influences decision-making; the impact of partial forms of normative adherence; and the relationship between the taboo and strategic narrative construction. The book provides a vital model for comprehending the taboo now in all these ways.

Fourth, the taboo can be discussed in relation to bioterrorism. This book does not engage with the issue of bioterrorism in that the threat was not raised as a major consideration during the historical period selected. The concern surrounding bioterrorism has intensified significantly since, however, and not least in relation to the 2001 post-9/11 anthrax mail strikes. It is argued here that the issue of bioterrorism cannot be addressed without an associated exploration of the bioweapons taboo. Bioterrorism must be considered in relation to the terms of, and motivations inherent to, the taboo that a terrorist may, or may not, be seen to break. The likelihood that terrorists will adhere to the same norms that restrain state actors has been queried, including specifically in relation to WMD (Hoffman 1997; Zanders 1999: 27–8; Walker 2010: 876). Breaking accepted norms may comprise an intrinsic aim of terrorism itself and terrorism can also disrupt the international expectations accepted by state actors, for example as the taboo against torture has been violated in response to the War on Terror (Schmidt and Sikkink 2019). Consequently, there exist key questions about whether and how the bioweapons taboo would hold under conditions of non-state consideration which require greater analytic attention. These are questions that the understanding laid out in this book can help answer. Analysis can also consider the ways state actors respond to the bioterrorist threat, where this again connects to the taboo. To draw once more upon Stern's conceptualization of bioterrorism as a type of dreaded risk, it can be shown that the issue is understood in disproportionate terms. This observation raises more questions concerning how this construction pertains to the taboo, not least how the taboo causes bioterrorism to become fetishized and how this perception then shapes political efforts to preclude terrorist use of bioweapons. Again, the book offers an understanding of the bioweapons taboo that can inform the answers to these crucial lines of enquiry.

Finally, the book underpins a set of lessons for policymakers that can be developed in the future and with respect to specific administrations and militaries. Most notably, the study demonstrates the critical fact that the taboo *is* a factor pertaining to biowarfare. Even if a given actor does not personally endorse the expectations of the taboo, normative prohibition is still relevant and relates to other actors that they engage with, domestically and internationally. Actors must then remain aware of the sheer opprobrium that would be attached to any action deemed to advocate for biowarfare. Even a hint of transgressive behaviour could attract extensive condemnation and sanction as an outcome of the taboo. In particular, actors must fully acknowledge the extent to which the taboo is codified within the BWC in that this agreement exacerbates expectations of normative compliance. Breaking the BWC would not 'just' comprise a breaking of the Convention. The act would constitute a breaking of the taboo, which would be considered a substantially more serious offence than simple non-compliance with an agreement's official terms due to the fetishized nature of the taboo. Doing so would cause uproar—perhaps not unlike the impassioned and vehement response to chemical weapons use in Syria or nuclear weapons development in North Korea. This book is then a vitally important reminder of what can happen if the taboo is transgressed or is perceived to have been transgressed, where this is then relevant to future policymaking.

The future of the bioweapons taboo

To the extent that bioweapons are constrained by normative proscription, could this change? Violation does have the potential to negate the taboo. Transgression, however, must necessarily be extensive or repeated in the form of a cascade to result in normative demise. Minor, one-off, or unproven allegations would likely be (and have so far been) insufficient to cause norm erosion. For example, *Biopreparat* did not trigger a domino effect or any decisive loss of faith in the taboo leading to the collapse of the BWC. Instead, evidence of violation has principally attracted widespread criticism, thereby reinforcing the taboo. The statement here is not to rule out the possibility of a violation that is capable of destabilizing the taboo. Yet contravention would have to occur on a transformative scale to do so and, at present, the taboo has consistently precluded such an event from happening.

Consequently, some other aspect pertaining to the current dynamics of the taboo would have to change in order to undermine the norm. The change could be technological. Biotechnology has undergone major scientific advances. For example, there is a significant concern pertaining to the attempted development of so-called 'designer' weapons which could be biologically engineered to attack only certain targets, such as by ethnicity (Wheelis 2004; Hart and Trapp 2012; Lewis 2016; Mackby 2018). In a situation where biological armaments have become more destructive, more targeted, or more efficient, this progress could engender shifts in the normative context that lead to the taboo's erosion. Changing what these weapons can do could change how we think about them.

Any shift in technological capacity, however, would still have to overcome the inherent aversion at the centre of the bioweapons taboo. Indeed, the act of effectively making the disease in question even more repellent via the scientific advancements identified could actually reinforce the norm as much as we might speculate that it could be undermined. The more extensive the contamination, the greater the sense of disgust and rejection. The advances highlighted here are already controversial in that they are perceived to relate to the manipulation of biology. Consider the opposition to genetic engineering, whereby 'playing god' with human life is itself deemed a taboo and an immoral act (Evans 2002). The same concern can be applied to future bioweapons, especially those that could target via genetic make-up. Furthermore, the BWC has been seen as actively slowing and even preventing scientific developments in respect of bioweapons. Lederberg (2001: 1071) said: 'Were it not for the Biological Weapons Convention, a gradually escalating technology race would have amplified even further this threat to human existence.' Consequently, it is not automatic that scientific advances will undermine the taboo. Indeed, the situation may be the other way around.

Yet it is also unlikely that the taboo will ever achieve full robustness in the sense that it becomes absolute. Expecting that there will never be any type of violation of the taboo going forward is optimistic at best; dangerous at worst. An acceptance that the taboo is imperfect, however, should not detract from the ways in which the norm has persisted. Unlike other weapons which are also deemed abhorrent but have still been employed—for example, chemical weapons—bioweapons use has not clearly comprised an active feature of modern conflict. Allegations of use have been made and state actors have chosen to develop these armaments. Yet there has been no major engagement of biowarfare tactics. Biowarfare remains an anathema. Events such as the Covid-19 pandemic serve only to reinforce the normative perception that disease is an intrinsic horror and, as such, that it should not be employed even as the means to secure. Bioweapons are recognized as a disgusting, immoral, and intolerable means to harm others. And this taboo sentiment is not likely to change anytime soon because it strikes at the core of our bodies and our very lives. For, as the author Thomas Mann (1999: 495) famously expressed it, 'all interest in disease and death is only another expression of interest in life'.

References

(1675) *Strasbourg Agreement.*

(1863) *General Orders No. 100 [Lieber Code].*

(1868) *Declaration Renouncing the Use, in Time of War, of Explosive Projectiles Under 400 Grammes Weight [St Petersburg Declaration].*

(1874) 'The Brussels Conference', *The New York Times*, 29 July: 4.

(1874) *Project of an International Declaration concerning the Laws and Customs of War [Brussels Declaration].*

(1899) 'Capt. Mahan's Argument', *New York Times*, 21 July: 7.

(1899) 'Close of the Hague Conference', *The Observer*, 30 July: 5.

(1899) *Convention II with Respect to the Laws and Customs of War on Land.*

(1899) 'Parting Reflections', *Manchester Guardian*, 1 August: 7–8.

(1899) 'The Peace Conference', *New York Times*, 23 June: 7.

(1899) 'The Peace Conference: Asphyxiating Shells and Expansive Bullets', *The Manchester Guardian*, 22 July: 7.

(1899) 'The Peace Conference: England and the Dum Dum Bullet', *Manchester Guardian*, 17 July: 7.

(1899) 'The Peace Conference: Private Property at Sea', *Manchester Guardian*, 23 June: 5.

(1899) 'Point Won at the Hague', *New York Times*, 18 July: 3.

(1905) 'Hague Declarations and Wishes', *International Law Studies Series* 5: 132–46.

(1907) 'On the Eve of Second Hague Conference', *New York Times*, 9 June: SM8.

(1915) 'Medicine, Magic, and Religion', *British Medical Journal* 2(2864): 751–3.

(1918) 'The Use of Asphyxiating Gas', *British Medical Journal* 2(3007): 168–9.

(1919) *Treaty of Peace with Germany [Treaty of Versailles].*

(1922) 'Popularity versus Security', *The Journal of Industrial and Engineering Chemistry* 14(3): 178.

(1922) *Treaty Relating to the Use of on Submarines and Noxious Gases in Warfare, [Washington Treaty].*

(1922) 'The War on Chemical Warfare', *The Journal of Industrial and Engineering Chemistry* 14(2): 98.

(1925) 'Agree on Search of Persian Gulf Ships', *New York Times*, 9 June: 6.

(1925) 'Arms Conference for Outlawing Gas', *New York Times*, 6 June: 1, 6.

(1925) 'Chemists Denounce Geneva Ban on Gas', *New York Times*, 5 August: 12.

(1925) 'Parley on War Gas Favored at Geneva', *New York Times*, 27 May: 14.

(1925) *Protocol for the Prohibition of the Use in War of Asphyxiating, Poisonous or Other Gases, and of Bacteriological Methods of Warfare [Geneva Protocol].*

(1926) 'Pershing Backs Poison Gas Ban', *New York Times*, 11 December: 3.

(1927) 'Our Position on Poison Gas', *New York Times*, 31 January: 5.

(1934) 'Germs as War Weapon: Paris and Mr. Wickham Steed', *The Observer*, 1 July: 18.

(1934) 'Germs in the War: A Frenchman Examines Their Possibilities as Weapons', *New York Times*, 28 October: XX6.

(1937) 'Bacterial Warfare', *New York Times*, 29 September: 22.

(1937) 'Germans Studying Germ War Tactics', *New York Times*, 22 September: 14.

(1941) 'Nazis Held Ready for Germ Warfare', *New York Times*, 17 September: 14.

(1942) 'Invisible Weapon: Plague from Planes over the Chinese City of Changteh', *Time* 39(10): 22.
(1945) 'United States—Great Britain—Canada: Declaration on Atomic Energy', *The American Journal of International Law; Supplement Official Documents* 40(1): 48–50.
(1946) 'Convicts Aid Germ Battle', *Oakland Tribune*, 5 January: 1.
(1946) 'Ounce of New Poison Could Kill Population of U.S. and Canada', *New York Times*, 19 September: 1.
(1946) 'Pandora's Box', *New York Times*, 12 January: 14.
(1947) 'Bacteria Put Above A-Bomb as Weapon', *New York Times*, 16 October: 20.
(1947) 'Merck Urges World to Bar All Weapons', *New York Times*, 8 November: 2.
(1949) 'Atomic Bomb Now "Child's Play": New Organic Weapon', *The Manchester Guardian*, 10 September: 7.
(1949) 'Minutes of the Eighth Meeting of the Policy Planning Staff on the International Control of Atomic Energy', *Foreign Relations of the United States, 1949, National Security Affairs, Foreign Economic Policy, Volume I*: 204.
(1950) 'Colorado Beetle: East German Story Becomes Involved', *The Manchester Guardian*, 8 June: 10.
(1950) 'Germ Warfare in Korea?', *Science* 58(2): 22.
(1950) 'Soviet Bug Charge "Phony," Says U.S.', *New York Times*, 3 July: 3.
(1951) 'Eleven Weeks of Bad Faith', *New York Times*, 12 September: 30.
(1951) 'Enemy Disease Smog Held a Possibility', *New York Times*, 24 October: 18.
(1951) 'National Intelligence Estimate', *Foreign Relations of the United States, 1951, Europe: Political and Economic Developments, Volume IV, Part 2*: 1622–9.
(1951) 'Relief Handicaps in Korea Stressed', *New York Times*, 22 May: 3.
(1951) 'Soviet Organ Sees Confusion in U.S.', *New York Times*, 3 April: 6.
(1952) 'Germs and Propaganda', *New York Times*, 29 March: 14.
(1952) 'Reds Call U.S. Atlas Germ Warfare Chart', *New York Times*, 3 October: 5.
(1952) 'Reds' Germ Charges Called "Black Lies"', *New York Times*, 7 June: 2.
(1952) 'Russians Demand U.S. Drop Atom Aim', *New York Times*, 9 May: 9.
(1952) 'U.N. Health Head Hits Germ Charge', *New York Times*, 4 April: 3.
(1952) 'U.S. May Urge U.N. Study "Germ War"', *New York Times*, 16 March: 3.
(1953) 'Russia Accused by U.S.: "Directed Torture of Airmen"', *Manchester Guardian*, 27 October: 1.
(1954) *Them!* G. Douglas, [film]
(1958) 'U.S. Seeks Arms That Do Not Kill', *New York Times*, 10 May: 8.
(1960) 'Army Conducts Nerve Gas Tests', *New York Times*, 28 August: 33.
(1960) 'Germ War Defense Displayed by Army', *New York Times*, 26 August: 3.
(1962) 'Research Memorandum Prepared in the Bureau of Intelligence and Research', *Foreign Relations of the United States, 1961–1963, Volume XXV, Organization of Foreign Policy; Information Policy; United Nations; Scientific Matters*: https://history.state.gov/historicaldocuments/frus1961-63v25/d219
(1965) 'Gas (Nonlethal) in Vietnam', *New York Times*, 24 March: 42.
(1966) '22 Scientists Bid Johnson Bar Chemical Weapons in Vietnam', *New York Times*, 20 September: 1, 3.
(1966) 'U. of Penn Shuns Germ War Pact', *New York Times*, 12 November: 6.
(1967) *Treaty on the Principles Governing the Activities of States in the Exploration and Use of Outer Space, including the Moon and Other Celestial Bodies [Outer Space Treaty]*
(1967) 'U.S. Said to Study Disease', *New York Times*, 14 January: 10.
(1968) 'British Working Paper on Microbiological Warfare', *Documents on Disarmament 1968*: 569–71.

(1968) 'Tie to Army Ended by Biology Society', *New York Times*, 9 May: 29.
(1968) *Treaty on the Non-Proliferation of Nuclear Weapons [Nuclear Non-Proliferation Treaty]*.
(1969) '... and Toxins', *New York Times*, 21 December: E14.
(1969) 'Analytical Summary Prepared by the National Security Council Staff', *Foreign Relations of the United States, 1969–176, Volume E-2, Documents on Arms Control and Nonproliferation, 1969–1972*: https://history.state.gov/historicaldocuments/frus1969-76ve02/d152
(1969) 'Authorization of Appropriations for Fiscal Year 1970 for Military Procurement, Research and Development, and for the Construction of Missile Test Facilities at Kwajalien Missile Range, and Reserve Component Strength', *Congressional Record – Senate*, 11 August: 23188–23232.
(1969) 'Briefing Book Prepared in the United States Information Agency', *Foreign Relations of the United States, 1917–1972, Volume VIII, Public Diplomacy, 1969–1972*: https://history.state.gov/historicaldocuments/frus1917-72PubDipv08/d1
(1969) 'Candor at the Pentagon', *New York Times*, 23 May: 46.
(1969) 'Defense's Dark Corner', *New York Times*, 23 April: 46.
(1969) 'Destroy this Weapon', *Christian Science Monitor*, 8 July: 18.
(1969) 'A Gas Leak Compels 200 to Evacuate', *New York Times*, 12 December: 33.
(1969) 'A Hard Look at Gas Warfare is Overdue', *The Washington Post*, 24 June: A18.
(1969) 'Laird Defends Chemical Weaponry', *New York Times*, 29 July: 4.
(1969) 'Laird Says Pentagon Opposed Nixon's Weapons Ban at First', *New York Times*, 2 December: 24.
(1969) 'Minutes of National Security Council Meeting', *Foreign Relations of the United States, 1969–1976, Volume XXXIV, National Security Policy, 1969–1972*: https://history.state.gov/historicaldocuments/frus1969-76v34/d103
(1969) 'Minutes of Review Group Meeting', *Foreign Relations of the United States, 1969–1976, Volume XXXIV, National Security Policy, 1969–192*: https://history.state.gov/historicaldocuments/frus1969-76v34/d97
(1969) 'Minutes of the Secretary of Defense's Staff Meeting', *Foreign Relations of the United States, 1969–176, Volume E-2, Documents on Arms Control and Nonproliferation, 1969–1972*: https://history.state.gov/historicaldocuments/frus1969-76ve02/d142
(1969) 'U.S. Chemical Arms a Deterrent, Laird Says,' *Los Angeles Times*, 29 July: A4.
(1969) 'War With Gas and Germ', *New York Times*, 7 April: 42.
(1969) 'Scientist Criticizes Germ War Research', *New York Times*, 30 June: 32.
(1970) 'Chemical and Biological Warfare', *Bulletin of the American Academy of Arts and Sciences* 23(4): 2–14.
(1970) 'General Kazimierz Sosnowski, the United States and the Ban on Bacteriological Weapons', *The Polish Review* 15(2): 105–13.
(1971) 'Brezhnev Speaks', *New York Times*, 31 March: 44.
(1971) 'Soviet Move to Ban Germ Warfare', *The Guardian*, 31 March: 3.
(1971) 'Telegram 72644 From the Department of State to the Mission in Geneva', *Foreign Relations of the United States, 1969–176, Volume E-2, Documents on Arms Control and Nonproliferation, 1969–1972*: https://history.state.gov/historicaldocuments/frus1969-76ve02/d226
(1971) *Treaty on the Prohibition of the Emplacement of Nuclear Weapons and Other Weapons of Mass Destruction on the Seabed and the Ocean Floor and in the Subsoil Thereof [Seabed Treaty]*.
(1971) 'US urges Russia to Drop Dual Request', *The Guardian*, 10 March: 4.
(1972) *Convention on the Prohibition of the Development, Production and Stockpiling of Bacteriological (Biological) and Toxin Weapons and on Their Destruction [Biological Weapons Convention]*.

(1972) 'Editorial Note', *Foreign Relations of the United States, 1969–176, Volume E-2, Documents on Arms Control and Nonproliferation, 1969–1972*: https://history.state.gov/historicaldocuments/frus1969-76ve02/d256
(1975) 'Editorial Note', *Foreign Relations of the United States, 1969–1976, Volume E-14, Part 2, Documents on Arms Control and Nonproliferation, 1973–1976*: https://history.state.gov/historicaldocuments/frus1969-76ve14p2/d114
(1975) 'Storage of Toxins Defended by Army', *New York Times*, 20 October: 16.
(1975) 'Subway Here Used to Assess Threat of a Germ Attack', *New York Times*, 17 September: 27.
(1979) *Alien*. R. Scott, [film]
(1980) 'Editorial Note', *Foreign Relations of the United States, 1977–1980, Volume XXVI, Arms Control and Nonproliferation*: https://history.state.gov/historicaldocuments/frus1977-80v26/d127
(1980) 'Memorandum of Conversation', *Foreign Relations of the United States, 1977–1980, Volume VI, Soviet Union*: https://history.state.gov/historicaldocuments/frus1977-80v06/d290
(1980) 'Summary of Conclusions of a Special Coordination Committee Meeting', *Foreign Relations of the United States, 1977–1980, Volume XXVI, Arms Control and Nonproliferation*: https://history.state.gov/historicaldocuments/frus1977-80v26/d112
(1989) *The Bible: New Revised Standard Version*: https://www.biblegateway.com/versions/New-Revised-Standard-Version-NRSV-Bible/
(1991) *Laws of Manu*. Translated by B. Smith and W. Doniger, London: Penguin.
(1993) *The Convention on the Prohibition of the Development, Production, Stockpiling and Use of Chemical Weapons and on their Destruction [Chemical Weapons Convention]*.
(2001) 'U.S. Rejection of Protocol to Biological Weapons Convention', *The American Journal of International Law* 95(4): 899–901.
(2002) 'Biological Weapons Convention—Comatose and barely alive', *Current Science* 82(1): 13–14.
(2005) *War Book of the General Staff*. Pennsylvania: Stackpole Books.
ACDA (1969) 'Eighth Annual Report of the United States Arms Control and Disarmament Agency', *Documents on Disarmament* 1968: 810–48.
Acheson, D. (1952a) 'The Pattern of Leadership—A Pattern of Responsibility', *Department of State Bulletin* XXVII(691): 423–7.
Acheson, D. (1952b) 'Recent Soviet Maneuvers', *Department of State Bulletin* XXVI(670): 666–7.
Acheson, D. (1952c) 'Refutation of Trumped-Up "Evidence" of Germ Warfare Charge', *Department of State Bulletin* XXVI(673): 777.
Acheson, D. (1952d) 'Further Denial of Soviet "Germ Warfare" Allegations', *Department of State Bulletin* XXVI(667): 529.
Acheson, D. (1952e) '"Germ Warfare" Charges Called Fabrication', *Department of State Bulletin* XXVI(664): 427–8.
Acheson, D. (1970) *Present at the Creation: My Years in the State Department*. London: Hamish Hamilton Ltd.
Addo, I.Y. (2020) 'Double pandemic: racial discrimination amid coronavirus disease 2019', *Social Sciences & Humanities Open* 2(1): 1–4.
Adler-Nissen, R. (2014) 'Stigma Management in International Relations: Transgressive Identities, Norms, and Order in International Society', *International Organization* 68(1): 143–76.
Aldrich, J.H. et al. (2006) 'Foreign Policy and the Electoral Connection', *Annual Review of Political Science* 9(1): 477–502.
Alibek, K. (1999) *Biohazard: The Chilling True Story of the Largest Covert Biological Weapons Program in the World—Told from Inside by the Man Who Ran It*. London: Hutchinson.

Allan, K. and Burridge, K. (2006) *Forbidden Words: Taboo and the Censoring of Language*. Cambridge: Cambridge University Press.
Allen, S.J. (2007) *An Analysis of Factors Leading to U.S. Renunciation of Biological Weapons* [PhD Thesis, George Mason University].
Alolaimy, M. (2017) 'Disarmament: Evolution or Revolution', *World Affairs* 21(1): 32–47.
Alt, J.E., Calvert, R.L., and Humes, B.D. (1988) 'Reputation and Hegemonic Stability: A Game-Theoretic Analysis', *The American Political Science Review* 82(2): 445–66.
Ampuja, M. (2020) 'The Blind Spots of Digital Innovation Fetishism', in Stocchetti, M. (ed.), *The Digital Ages and Its Discontents: Critical Reflections in Education*. Helsinki: Helsinki University Press, 31–53.
Anderson, W. (1999) 'Disease and Its Meanings', *Health and History* 1(4): 245–9.
Angyal, A. (1941) 'Disgust and Related Aversions', *Journal of Abnormal and Social Psychology* 36(3): 393–412.
Apollodorus (2008) *The Library of Greek Mythology*. Translated by Hard, R. Oxford: Oxford University Press.
Arneson, R.G. (1989) 'Oral History Interview with R. Gordon Arneson', *Truman Presidential Library*: https://www.trumanlibrary.gov/library/oral-histories/arneson
Arneson, R.J. (2007) 'Shame, Stigma, and Disgust in the Decent Society', *The Journal of Ethics* 11(1): 31–63.
Arnold, D. (1993) *Colonising the Body: State, Medicine and Epidemic Disease in Nineteenth-Century India*. Berkley: University of California Press.
Atkinson, C. (2010) 'Using Nuclear Weapons', *Review of International Studies* 36(4): 839–51.
Atlay, J.B. (1905) 'Legitimate and Illegitimate Modes of Warfare', *Journal of the Society of Comparative Legislation* 6(1): 10–21.
Auchter, J. (2017) 'On Viewing: The Politics of Looking at the Corpse', *Global Discourse* 7(2–3): 223–38.
Avdeyeva, O.A. (2012) 'Does Reputation Matter for States' Compliance with International Treaties? States Enforcement of Anti-Trafficking Norms', *The International Journal of Human Rights* 16(2): 298–320.
Bailey, K.C. (1995) 'Responding to the Threat of Biological Weapons', *Security Dialogue* 26(4): 383–97.
Baldwin, H.W. (1946a) 'New Weapons—Menace and Warning', *New York Times*, 30 June: 5, 51–3.
Baldwin, H.W. (1946b) 'Biological Warfare: Destruction by Disease Looms as Rival to Atomic Bomb and Gas', *New York Times*, 14 July: 4.
Baldwin, H.W. (1952) 'Korea Move Up to Soviet: Truce Developments Indicate Moscow May be Mapping World "Package Deal"', *New York Times*, 11 April: 3.
Balmer, B. (2002a) 'Biological Warfare: The Threat in Historical Perspective', *Medicine, Conflict and Survival* 18(2): 120–37.
Balmer, B. (2002b) 'Killing "Without the Distressing Preliminaries": Scientists' Defence of the British Biological Warfare Programme', *Minerva* 40(1): 57–75.
Barghoorn, F.C. (1954) 'The Soviet Image of the United States: A Deliberately Distorted Image', *The Annals of the American Academy of Political and Social Science* 295: 42–51.
Barnaby, W. (1997) 'Biological Weapons: An Increasing Threat', *Medicine, Conflict and Survival* 13(4): 301–13.
Barras, V. and Greub, G. (2014) 'History of Biological Warfare and Bioterrorism', *Clinical Microbiology and Infection* 20(6): 497–502.
Barrett, S. (1999) 'International Cooperation and the International Commons', *Duke Environmental Law and Policy Forum* 10: 131–46.

Bashford, A. (2006) 'Global Biopolitics and the History of World Health', *History of the Human Sciences* 19(1): 67–88.

Beard, J.M. (2007) 'The Shortcomings of Indeterminacy in Arms Control Regimes: The Case of the Biological Weapons Convention', *The Americans Journal of International Law* 101(2): 271–321.

Behre, A.A. (2005) 'Politicizing Indiscriminate Terror: Imagining an Inclusive Framework for the Anti-Landmines Movement', *The Journal of Environmental Development* 14(3): 375–93.

Ben Ouagrham-Gormley, S. (2013) 'Dissuading Biological Weapons Proliferation', *Contemporary Security Policy* 34(3): 473–500.

Benedictow, O. (2004) *The Black Death 1346–1353: The Complete History*. Woodbridge: The Boydell Press.

Bennett, I. (1970) 'The Significance of Chemical and Biological Warfare for the People', *Proceedings of the National Academy of Sciences of the United States of America* 65(1): 271–9.

Bentley, M. (2014) *Weapons of Mass Destruction and US Foreign Policy: The Strategic Use of a Concept*. London: Routledge.

Bentley, M. (2016) *Syria and the Chemical Weapons Taboo: Exploiting the Forbidden*. Manchester: Manchester University Press.

Bentley, M. (2018) 'Fetishised Data: Counterterrorism, Drone Warfare and Pilot Testimony', *Critical Studies on Terrorism* 11(1): 88–100.

Bentley, M. (2019a) 'The US has a History of Testing Biological Weapons on the Public—Were Infected Ticks Used too?', *The Conversation*, 22 July: https://theconversation.com/the-us-has-a-history-of-testing-biological-weapons-on-the-public-were-infected-ticks-used-too-120638

Bentley, M. (2019b) 'Agent Orange', in Fee, C.R. and Webb, J.B. (eds), *Conspiracies and Conspiracy Theories in American History: Volume I*. Santa Barbara: ABC-CLIO, 46–50.

Bentley, M. (2022) 'A New Model of "Taboo": Disgust, Stigmatization, and Fetishization', *International Studies Review*: 24(3), viac028: 1–25.

Bering, J. (2013) 'That's Disgusting', *Scientific American Mind* 24(5): 23–4.

Bernstein, B.J. (1987) 'The Birth of the U.S. Biological-Warfare Program', *Scientific American* 256(6): 116–21.

Betts, R.K. (1998) 'The New Threat of Mass Destruction', *Foreign Affairs*, 77(1): 26–41.

Bially Mattern, J. (2005) *Ordering International Politics: Identity, Crisis, and Representational Force*. London: Routledge.

Bicchieri, C. (2017) *Norms in the Wild: How to Diagnose, Measure, and Change Social Norms*. Oxford: Oxford University Press.

Bidwai, P. (1999) 'Towards a New Paradigm: Comprehensive Security, Weapons of Mass Destruction, and Biological Weapons', *Politics and the Life Sciences* 18(1): 109–11.

Binder, D. (1978) 'Bid to Mislead Nixon Laid to Soviet Spies', *New York Times*, 4 June: 18.

Bishop, H.G. (1934) 'Terror From the Sky: Reality of Bogy?', *New York Times*, 25 March: 4–5, 19.

Bjorkdahl, A. (2002) 'Norms in International Relations: Some Conceptual and Methodological Reflections', *Cambridge Review of International Affairs* 15(1): 9–23.

Black, I. (2003) 'Budget Clash Hits Talks on EU Constitution', *The Guardian*, 29 November: https://www.theguardian.com/world/2003/nov/29/eu.politics

Bleiker, R. and Hutchinson, E. (2008) 'Fear No More: Emotions and World Politics', *Review of International Studies* 34(S1): 115–35.

Boer, R. (2010) 'That Hideous Pagan Idol: Marx, Fetishism and Graven Images', *Critique* 38(1): 93–116.

Boffey, P.M. (1968) 'Nerve Gas: Dugway Accident Linked to Utah Sheep Kill', *Science* 162(3861): 1460–4.

Boivin, M. (1973) 'A Positive Approach to Taboo', *New Blackfriars* 54(634): 100–7.

Borrie, J. (2006) 'The Limits of Modest Progress: The Rise, Fall, and Return of Efforts to Strengthen the Biological Weapons Convention', *Arms Control Today* 36(8): 18–22.

Boyle, F.A. (1985) 'The Legal Distortions behind the Reagan Administration's Chemical and Biological Warfare Buildup', *Saint Louis University Law Journal* 30: 1175–94.

Braun, B. (2013) 'Power over Life: Biosecurity as Biopolitics', in Dobson, A., Barker, K., and Taylor, S.L. (eds), *Biosecurity: The socio-Politics of Invasive Species and Infectious Diseases*. London: Routledge, 45–55.

Brody, H. et al. (2014) 'United States Responses to Japanese Wartime Inhuman Experimentation after World War II: National Security and Wartime Exigency', *Cambridge Quarterly of Healthcare Ethics* 23(2): 220–30.

Brophy, L.P., Miles, W.D., and Cochrane, R.C. (1959) *The Chemical Warfare Service: From Laboratory to Field*. Washington, DC: Office of the Chief of Military History, United States Army.

Brungs, B.J. (1964) 'The Status of Biological Warfare in International Law', *Military Law Review* 24: 47–95.

Budryk, Z. (2020) 'Bolton Worried about Biological Weapons, NATO Withdrawal if Trump is Releected', *The Hill*, 23 June: https://thehill.com/homenews/administration/504022-bolton-worried-about-biological-weapons-nato-withdrawal-if-trump-is

Buell, R.L. (1922) *The Washington Conference*. New York: D. Appleton and Company.

Bujard, D. (1974) 'The Geneva Convention of 1864 and the Brussels Conference of 1874', *International Review of the Red Cross* 14(163): 527–37.

Bull, H. (1966) 'International Theory: The Case for a Classical Approach', *World Politics* 18(3): 361–77.

Bunn, G. (1969) 'Banning Poison Gas and Germ Warfare: Should the United States Agree?', *Wisconsin Law Review* 1969(2): 375–420.

Bunn, G. (1970) 'The Banning of Poison Gas and Germ Warfare: The U.N. Role', *The American Journal of International Law* 64(4): 194–9.

Burridge, K. (2004) *Blooming English: Observations on the Roots, Cultivation and Hybrids of the English Language*. Cambridge: Cambridge University Press.

Burriss, E.E. (1929) 'The Nature of Taboo and Its Survival in Roman Life', *Classical Philology* 24(2): 142–63.

Burt, R.R. (1982) 'Use of Chemical Weapons in Asia', *Department of State Bulletin* 82(2058): 52–4.

Burton, T. (1925) 'The Chairman of the American Delegation (Burton) to the Secretary of State', *Foreign Relations of the United States, 1925, Volume I*: https://history.state.gov/historicaldocuments/frus1925v01/d41

Bush, G.H.W. (1972) 'U.N. Commends Biological Weapons Convention and Requests Continued Negotiations on Prohibition of Chemical Weapons', *Department of State Bulletin* LXVI(1700): 102–7.

Bush, V. (1970) *Pieces of the Action*, London: Cassell & Company Ltd.

Buxton, G.E. (1943) 'Memorandum for the Secretary, Joint Chiefs of Staff; Subject: Implications of Recent Intelligence Regarding Alleged German Secret Weapon', in CIA 'OSS—casualties from Allied phosphorous bombing of Berlin and Hamburg; potential for chemical and biological warfare; activities of censorship office; China assessments; 1942–95': https://www.cia.gov/library/readingroom/document/cia-rdp13x00001r000100430005-2

Byfield, R.S. (1951) 'The Cold facts versus the Big Lie, the Glittering Promise, and the Half-Truth', *The Analysts Journal* 7(3): 28–30.

Carnahan, B.M. (1998) 'Lincoln, Lieber and the Laws of War: The Origins and Limits of the Principle of Military Necessity', *The American Journal of International Law* 92(2): 213–31.

Carpenter, C. and Montgomery, A.H. (2020) 'The Stopping Power of Norms: Saturation Bombing, Civilian Immunity, and U.S. Attitudes towards the Laws of War', *International Security* 45(2): 140–69.

Carus, W.S. (2017a) *A Short History of Biological Warfare: From Pre-History to the 21st Century. Centre for the Study of Weapons of Mass Destruction, Occasional Paper 12*. Washington, DC: National Defense University Press.

Carus, W.S. (2017b) 'A Century of Biological-Weapons Programs (1915–2015): Reviewing the Evidence', *Nonproliferation Review* 24(1–2): 129–53.

Cassell, P.G. (1983) 'Establishing Violations of International Law: "Yellow Rain" and the Treaties Regulating Chemical and Biological Warfare', *Stanford Law Review* 35(2): 259–95.

Cavanagh, C.J. (2014) 'Biopolitics, Environmental Change, and Development Studies', *Forum for Development Studies* 41(2): 273–94.

Chayes, A. and Chayes, A. (1995) *The New Sovereignty: Compliance with International Regulatory Agreements*. Cambridge, MA: Harvard University Press.

Chen, S. (2009) 'History of Three Mobilizations: A Reexamination of the Chinese Biological Warfare Allegations against the United States in the Korean War', *The Journal of American-East Asian Relations* 16(30): 213–47.

Chevrier, M.I. (1990) 'Verifying the Unverifiable: Lessons from the Biological Weapons Convention', *Politics and the Life Sciences* 9(1): 93–105.

Christopher, G.W. et al. (2001) 'Biological Warfare: A Historical Perspective', in Lederberg, J. (ed.), *Biological Weapons: Limiting the Threat*. Cambridge, MA: Belfer Center for Science and International Affairs, 17–35.

Christopher, W. (1980) 'Telegram From the Department of State to the Embassy in the Soviet Union', *Foreign Relations of the United States, 1977–1980, Volume XXVI, Arms Control and Nonproliferation*: https://history.state.gov/historicaldocuments/frus1977-80v26/d131

CIA (1944) 'Evaluation of Damage to be Expected of Biological Warfare', in CIA 'OSS—casualties from Allied phosphorous bombing of Berlin and Hamburg; potential for chemical and biological warfare; activities of censorship office; China assessments; 1942–95': https://www.cia.gov/library/readingroom/document/cia-rdp13x00001r000100430005-2

CIA (1947) 'Review of the World Situation as it Relates to the Security of the United States': https://www.cia.gov/library/readingroom/document/cia-rdp67-00059a000200200046-7

CIA (1948) 'Intelligence Memorandum No. 105: Soviet Capabilities to Wage Biological Warfare': https://www.cia.gov/library/readingroom/document/cia-rdp78-01617a000300060002-7

CIA (1949a) 'Intelligence Memorandum No. 226: Estimate of Status of Biological Warfare in the USSR (Category No. 6)': https://www.cia.gov/library/readingroom/document/cia-rdp78-01617a000700170001-2

CIA (1949b) 'Intelligence Memorandum No. 163: Foreign Activities in the Field of Biological Warfare (BW)': https://www.cia.gov/library/readingroom/document/0001134404

CIA (1950) 'Joint Periodic Intelligence Report on Soviet Intentions and Activities': https://www.cia.gov/library/readingroom/docs/CIA-RDP91T01172R000400200007-0.pdf

CIA (1951a) 'Chinese Communist Press Charges US Uses Bacteriological Weapons in Korea': https://www.cia.gov/library/readingroom/docs/CIA-RDP80-00809A000600390700-1.pdf

CIA (1951b) 'Says US General Decorated for Using Bacteriological Weapons': https://www.cia.gov/library/readingroom/docs/CIA-RDP80-00809A000600400441-7.pdf

CIA (1951c) 'The Prevention of Typhus Outbreaks': https://www.cia.gov/library/readingroom/docs/CIA-RDP80-00809A000600400331-9.pdf

CIA (1951d) 'Chinese Communist Propaganda Regarding Bacteriological Warfare': https://www.cia.gov/library/readingroom/docs/CIA-RDP82-00457R009400050010-6.pdf

CIA (1952a) 'Special Estimate: Communist Charges of US Use of Biological Warfare': https://www.cia.gov/library/readingroom/docs/CIA-RDP79S01011A000600050006-3.pdf
CIA (1952b) 'Suggested Countermeasures in the Case of Chinese Communist Charges of U.S. Bacteriological Warfare in Northeast China' https://www.cia.gov/library/readingroom/docs/CIA-RDP79S01011A000600050020-7.pdf
CIA (1952c) 'Communist Propaganda Themes': https://www.cia.gov/library/readingroom/docs/CIA-RDP82-00457R013000240003-2.pdf
CIA (1952d) 'Intelligence Memorandum: Intensified "peace campaign: gives no indication of Soviet policy change': https://www.cia.gov/library/readingroom/docs/CIA-RDP91T01172R000300300007-0.pdf
CIA (1953) 'Outline: Soviet reaction to US reports on BW and atrocities': https://www.cia.gov/library/readingroom/docs/CIA-RDP79R00890A000200010012-1.pdf
CIA (1957) 'Marshal Zhukov on the Nature of a Future War': https://www.cia.gov/library/readingroom/docs/1957-04-05.pdf
CIA (1969a) 'Intelligence Report Prepared by Directorate of Intelligence, Central Intelligence Agency', *Foreign Relations of the United States, 1969–176, Volume E-2, Documents on Arms Control and Nonproliferation, 1969–1972*: https://history.state.gov/historicaldocuments/frus1969-76ve02/d144
CIA (1969b) 'Soviet Chemical and Biological Warfare Capabilities': https://www.cia.gov/library/readingroom/docs/DOC_0000283815.pdf
CIA (1970) 'Central Intelligence Bulletin: 17 October': https://www.cia.gov/library/readingroom/docs/CIA-RDP79T00975A017400030001-5.pdf
CIA (1971a) 'Central Intelligence Bulletin: 30 March': https://www.cia.gov/library/readingroom/docs/CIA-RDP79T00975A018600090001-6.pdf
CIA (1971b) 'Central Intelligence Bulletin: 18 May': https://www.cia.gov/library/readingroom/docs/CIA-RDP79T00975A019000100001-9.pdf
Cicero, M.T. (1942) *De Oratore*. London: William Heinemann.
Cicero, M.T. (2014) *De Officiis*. Translated by Miller, W. Project Gutenberg: http://www.gutenberg.org/ebooks/47001
Clare, J. and Danilovic, V. (2010) 'Multiple Audiences in Reputation Building in International Conflicts', *Journal of Conflict Resolution* 54(6): 869–82.
Clark, M. (1954) *From the Danube to the Yalu*. London: George G. Harrup.
Clarke, R. (1968) *The Silent Weapons: The Realities of Chemical and Biological Warfare*. New York: David McKay Company Inc.
Clifford, C.M. (1971) 'Oral History Interview with Clark M. Clifford', *Truman Presidential Library*: https://www.trumanlibrary.gov/library/oral-histories/cliford2
Clubb, O.E. (1950) 'Memorandum by the Director of the Office of Chinese Affairs (Clubb) to the Assistant Secretary of State for Far Eastern Affairs (Rusk)', *Foreign Relations of the United States, 1950, Korea, Volume VII*: 1087–92.
Cochrane, R. (1947) History of Chemical Warfare Service in World War II (1 July 1940–15 August 1945), Volume II: Biological Warfare Research. Historical Section, Plans, Training and Intelligence Division, Office of Chief, Chemical Corps.
Cohen, B.V. (1952a) 'The Case against Unilateral Disarmament', *Department of State Bulletin* XXVI(679): 1029–31.
Cohen, B.V. (1952b) 'U.S. Views on Dealing with Germ Warfare Elimination as a Separate Problem', *Department of State Bulletin* XXVII(687): 294–7.
Cohen, B.V. (1952c) 'Advancement of World Peace through Disarmament', *Department of State Bulletin* XXVI(666): 501–9.
Cohen, B.V. (1952d) 'U.S. Charts Course Towards Comprehensive, Coordinated Disarmament Program', *Department of State Bulletin* XXVI(672): 752–5.

Cohen, B.V. (1952e) 'U.S. Position on Germ Warfare', *Department of State Bulletin* XXVI(676): 911–13.

Cohen, B.V. (1953) 'Retiring Deputy U.S. Representative on Disarmament Commission Reports to the President', *The Department of State Bulletin*, XXVIII(709): 142–54.

Cohen, S.K. (2012) 'Pandemics: Waves of Disease, Waves of Hate from the Plague of Athens to A.I.D.S.', *Historical Research* 85(230): 535–55.

Cole, L.A. (1997) *The Eleventh Plague: The Politics of Biological and Chemical Warfare*. New York: W.H. Freeman and Company.

Cole, L.A. (1998) 'The Poison Weapons Taboo: Biology, Culture, and Policy', *Politics and the Life Sciences* 17(2): 119–32.

Comor, E. (2017) 'Technological Fetishism and US Foreign Policy: The Mediating Role of Digital ICTs', *The Political Economy of Communication* 5(2): 3–21.

Compton, W. (1952a) 'Mutual Security Requires Mutual Understanding', *Department of State Bulletin* XXVI(670): 668–72.

Compton, W. (1952b) 'The Voice of America at the Water's Edge', *Department of State Bulletin* XXVI(675): 864–7.

Conway, P.G. (1972) *An Analysis of Decision Making on U.S. Chemical and Biological Warfare Policies in 1969* [PhD Thesis, Purdue University].

Cook, J. (1967) *The Journals of Captain James Cook on his Voyages of Discovery: Volume III, Part I*. Cambridge: Cambridge University Press.

Cook, J. (2014) *Captain Cook's Journal during his First Voyage round the World, made in H.M. Bark* Endeavour, *1768–71*. Cambridge: Cambridge University Press.

Cook, T. (2000) '"Against God-Inspired Conscience": The Perception of Gas Warfare as a Weapon of Mass Destruction', *War & Society* 18(1): 47–69.

Cooper, N. (2011) 'Humanitarian Arms Control and Processes of Securitization: Moving Weapons along the Security Continuum', *Contemporary Security Policy* 32(1): 134–58.

Cortell, A.P. and Davis, J.W. (2005) 'When Norms Clash: International Norms, Domestic Practices, and Japan's Internalisation of the GATT/WTO', *Review of International Studies* 31(1): 3–25.

Cortesi, A. (1952) 'Naples Welcomes Ridgway Warmly', *New York Times*, 18 June: 5.

Cowdrey, A.E. (1984) '"Germ Warfare" and Public Health in the Korean Conflict', *Journal of the History of Medicine and Allied Sciences* 39(20): 153–72.

Cowhey, P.F. (1993) 'Domestic Institutions and the Credibility of International Commitments: Japan and the United States', *International Organization* 47(2): 299–326.

Coyle, P.E. (2003) 'Weapons of Mass Destruction in Iraq, North Korea and Iran: Hype, Hope or Hysteria?', Physics and Society, October: 10–15.

Crane, C. (2001) 'Chemical and Biological Warfare During the Korea War: Rhetoric and Reality', *Asian Perspective* 25(3): 61–83.

Crawford, N. (2013) 'Emotions and International Security: *Cave! Hic Libido*', *Critical Studies on Security* 1(1): 121–3.

Creasy, W. (1950) 'Presentation to the Secretary of Defense's Ad Hoc Committee on CEBAR', Record Number 54874: Chemical and Biological Warfare Collection, *National Security Archive*.

Crescenzi, M.J.C. (2007) 'Reputation and Interstate Conflict', *American Journal of Political Science* 51(2): 382–96.

Crescenzi, M.J.C., Kathman, J.D., and Long, S.B. (2007) 'Reputation, History, and War', *Journal of Peace Research* 44(6): 651–67.

Crescenzi, M.J.C. et al. (2012) 'Reliability, Reputation, and Alliance Formation', *International Studies Quarterly* 56(2): 259–74.

Crichton, M. (1995) *The Andromeda Strain*. London: Arrow Books.

Cross, G. and Klotz, L. (2020) 'Twenty-First Century Perspectives on the Biological Weapon Convention: Continued Relevance or Toothless Paper Tiger', *Bulletin of the Atomic Scientists* 76(4): 185–91.

Curtis, V. (2007) 'Dirt, Disgust and Disease: A Natural History of Hygiene', *Journal of Epidemiology and Community Health* 61(8): 660–4.

Curtis, V. (2011) 'Why Disgust Matters', *Philosophical Transactions: Biological Sciences* 366(1583): 3478–90.

Curtis, V., Aunger, R., and Rabie, T. (2004) 'Evidence that Disgust Evolved to Protect from Risk of Disease', *Proceedings: Biological Sciences* 271: S131–3.

Curtis, V. and Biran, A. (2001) 'Dirt, Disgust, and Disease: Is Hygiene in Our Genes?', *Perspectives in Biology and Medicine* 44(1): 17–31.

CWS (1939) *Technical Study No. 10: Bacteriological Warfare Possibilities*. Washington, DC: Chemical Warfare Service.

D.S. (1973) 'Chemical Corps To Be Diluted', *Science* 179(4074): 664.

Dallen, D.J. (1952) 'Soviet Plans for Germ War: Charges against U.N. Said to Follow Usual Moscow Pattern', *New York Times*, 15 May: 30.

Dando, M. (1997) 'Development of International Legal Constraints on Biological Warfare in the 20th Century', *The Finnish Yearbook of International Law* 8: 1–69.

Dando, M. (2005) 'The Bioterrorist Cookbook', *The Bulletin of the Atomic Scientists*, 61(6): 34–9.

Danzig, R. and Berkowsky, P.B. (2001) 'Why Should We Be Concerned about Biological Warfare?' in Lederberg, J. (ed.), *Biological Weapons: Limiting the Threat*. Cambridge, MA: Belfer Center for Science and International Affairs, 9–14.

Darwin, C. (2007) *The Expression of the Emotions in Man and Animals*. New York: Dover Publications Inc.

Davey, G.C.L. (2011) 'Disgust: The Disease-Avoidance Emotion and its Dysfunctions', *Philosophical Transactions: Biological Sciences* 366(1583): 3453–65.

Davis, G.B. (1907) 'Doctor Francis Lieber's Instructions for the Government of Armies in the Field', *The American Journal of International Law* 1(1): 13–25.

Davis Gibbons, R, and Lieber, K. (2019) 'How Durable is the Nuclear Weapons Taboo?', *Journal of Strategic Studies* 42(1): 29–54.

Dawson, A. (1967) 'Hannibal and Chemical Warfare', *The Classical Journal* 63(3): 117–25.

Reinke de Buitrago, S. (2016) 'The Role of Emotions in US Security Policy towards Iran', *Global Affairs* 2(2): 155–64.

De Palma, S. (1969) 'What the United Nations Means to America—Now', *Department of State Bulletin* LXI(1584): 374–7.

Deaux, G. (1969) *The Black Death, 1347*. London: Hamilton.

DeForest, P.H. (1990) 'The Matrix of Biological Disarmament: Strategy, Law, and Technology', *Politics and the Life Sciences* 9(1): 21–35.

Dehority, W. (2020) 'Infectious Disease Outbreaks, Pandemics, and Hollywood—Hope and Fear across a Century of Cinema', *JAMA* 323(19): 1878–80.

Deitelhoff, N. and Zimmermann, L. (2020) 'Things We Lost in the Fire: How Different Types of Contestations Affect the Robustness of International Norms', *International Studies Review* 22(1): 51–76.

DeLupis, I.D. (1987) *The Law of War*. Cambridge: Cambridge University Press.

Dembek, Z.F. (2006) 'The History and Threat of Biological Weapons and Bioterrorism', in McIsaac, J.H. (ed.), *Preparing Hospitals for Bioterror: A Medical and Biomedical Systems Approach*. Cambridge, MA: Elsevier Academic Press, 17–35.

Devlin, P. (1968) *The Enforcement of Morals*. Oxford: Oxford University Press.

Di Cicco, C. (2014) *Bioterrorism*. Scotts Valley: CreateSpace.

Dimitriu, G. R. (2011) 'Winning the Story War: Strategic Communication and the Conflict in Afghanistan', *Public Relations Review* 38(2): 195–207.

Ditrich, H. (2017) 'The Transmission of the Black Death to Western Europe: A Critical Review of the Existing Evidence', *Mediterranean Historical Review* 32(1): 25–39.

DoA (1956) *The Law of Land Warfare*.

DoA (1976) *The Law of Land Warfare*.

DoA (1977) *US Army Activity in the U.S. Biological Warfare Programs: Volume 1*.

DoD (1969) 'Talking Paper Prepared in the Department of Defense', *Control and Nonproliferation, 1969–1972*: https://history.state.gov/historicaldocuments/frus1969-76ve02/d168

Dolan, T.M. (2013) 'Unthinkable and Tragic: The Psychology of Weapons Taboos in War', *International Organization* 67(1): 37–63.

DoS (1952a) 'United States Policy Towards the USSR and International Communism: A Monthly Digest of Pertinent Data Appearing in Department of State Publications.' Department of State, Division of Historical Policy Research, Office of Public Affairs.

DoS (1952b) 'Memorandum of the Substance of Discussion at a Department of State Joint Chiefs of Staff Meeting', *Foreign Relations of the United States, 1952–1954, Korea, Volume VX, Part I*: 82–6.

DoS (1952c) 'U.S. Facing Renewed Communist Germ Warfare Charges', *Department of State Bulletin* XXVII(692): 475–6.

DoS (1952d) 'Red Cross to Investigate "Germ Warfare" Charges', *Department of State Bulletin* XXVI(665): 452–3.

DoS (1952e) 'The United States in the United Nations: March 15–March 27, 1952', *Department of State Bulletin* XXVI(666): 515–16.

DoS (1952f) 'The United States in the United Nations: June 20–July 3, 1952', *Department of State Bulletin* XXVII(680): 38–9.

DoS (1952g) 'Senate Begins Consideration of German Contractual Agreements and North Atlantic Treaty Protocol', *Department of State Bulletin* XXVI(678): 971–9.

DoS (1952h) 'Intelligence Brief: Chou En-Lai and Pei-P'ing's "People's Daily" Amplify Charges of US Bacteriological Warfare Against Manchuria': https://www.cia.gov/library/readingroom/docs/CIA-RDP79S01011A000600050001-8.pdf

DoS (1952i) 'The United States in the United Nations: June 6–June 19, 1952', *Department of State Bulletin* XXVI(679): 1041–2.

DoS (1982) Special Report No. 104, Chemical Warfare in Southeast Asia and Afghanistan: An Update: https://www.cia.gov/readingroom/docs/CIA-RDP87R00029R000400790001-9.pdf

DoS (1986) 'Cable No. 291907, Secretary of State to U.S. Mission, Geneva: Visit to Washington of Professor Meselson, Secret', *National Security Archive*: https://nsarchive2.gwu.edu/NSAEBB/NSAEBB61/Sverd27.pdf

Douglas, M. (1984) *Purity and Danger: An Analysis of the Concepts of Pollution and Taboo*. London: Routledge.

Douglass, J.D. and Livingstone, N.C. (1987) *America the Vulnerable*. Lexington: Heath.

Drobysz, S. (2021) 'Verification and Implementation of the Biological and Toxin Weapons Convention', *Nonproliferation Review* 27(4–6): 487–97.

DuBridge, L.A. (1969) 'Memorandum for Dr. Henry A. Kissinger', *National Security Archive*: http://nsarchive2.gwu.edu.946elmp01.blackmesh.com/NSAEBB/NSAEBB58/RNCBW5.pdf

Dulles, A.W. (1955) 'Soviet Strategic Concepts': https://www.cia.gov/library/readingroom/docs/CIA-RDP80B01676R001100160004-7.pdf

Dyson, F. (1984) *Weapons and Hope*. New York: Harper and Row.

Eagleburger, L.S. (1982) 'Preserving Western Independence and Security', *Department of State Bulletin* 82(2058): 36–40.

Eagleburger, L.S. (1983) 'Yellow Rain: The Arms Control Implications', *Department of State Bulletin* 83(2073): 77–8.

Earle, R. (1980) 'Memorandum From the Director of the Arms Control and Disarmament Agency (Earle) to the President's Assistant for National Security Affairs (Brzezinski)', *Foreign Relations of the United States, 1977–80, Volume XXVI, Arms Control and Nonproliferation*: https://history.state.gov/historicaldocuments/frus1977-80v26/d109

Eden, L. 2010. 'The Contingent Taboo', *Review of International Studies* 36(4): 831–7.

Eisenhower, D.D. (1946) 'The Chief of Staff of the United States Army (Eisenhower) to the United States Representative on the Atomic Energy Commission (Baruch)', *Foreign Relations of the United States, 1946, General: The United Nations, Volume I*: 854–7.

Eisenhower, D.D. (1960a) 'The President's News Conference', *The American Presidency Project*: https://www.presidency.ucsb.edu/documents/the-presidents-news-conference-219

Eisenhower, D.D. (1960b) 'Address by President Dwight Eisenhower to the UN General Assembly', *DoS*: https://2009-2017.state.gov/p/io/potusunga/207330.htm

Elias, N. (2000) *The Civilizing Process*. Oxford: Blackwell.

Ellen, R. (1988) 'Fetishism', *Man* 23(2): 213–35.

Elster, J. (1989) *The Cement of Society: A Study of Social Order*. Cambridge: Cambridge University Press.

Elster, J. (1999) *Alchemies of the Mind: Rationality and the Emotions*. Cambridge: Cambridge University Press.

Endicott, S. and Hagerman, E. (1998) *The United States and Biological Warfare: Secrets from the Early Cold War and Korea*. Bloomington: Indiana University Press.

Enemark, C. (2007) 'Health Security Challenges: Biological Weapons and Pandemic Influenza', *Security Challenges* 3(3): 131–43.

Enia, J. and Fields, J. (2014) 'The Relative Efficacy of the Biological and Chemical Weapon Regimes', *Nonproliferation Review* 21(1): 43–64.

Epstein, C. (2008) *The Power of Words in International Relations: Birth if an Anti-Whaling Discourse*. Cambridge, MA: MIT Press.

Erskine, T. and Lebow, R.N. (2012) *Tragedy and International Relations*. London: Palgrave.

Evans, J.H. (2002) *Playing God? Human Genetic Engineering and the Rationalization of Public Bioethical Debate*. Chicago: University of Chicago Press.

Ewin, T. (2020) 'Modern resonances of Imperial Germany's biological-Warfare Sabotage Campaign, 1915–18', *Nonproliferation Review* 27(4–6): 277–87.

Fattah, K. and Fierke, K.M. (2009) 'A Clash of Emotions: The Politics of Humiliation and Political Violence in the Middle East', *European Journal of International Relations* 15(1): 67–93.

Feinberg, J. (1988) *The Moral Limits of the Criminal Law Volume IV: Harmless Wrongdoing*. Oxford: Oxford University Press.

Fellmeth, A.X. and Horwitz, M. (2009) *Guide to Latin in International Law*. Oxford: Oxford University Press.

Fellows, L. (1969) 'Bonn Urges Ban on Poison Arms', *New York Times*, 13 September: 7.

Fershtman, C., Gneezy, U., and Hoffman, M. (2011) 'Taboos and Identity: Considering the Unthinkable', *American Economic Journal: Microeconomics* 3(2): 139–64.

Fidler, D.P. and Gostin, L.O. (2007) *Biosecurity in the Global Age: Biological Weapons, Public Health, and the Rule of Law*. Stanford: Stanford University Press.

Findlay, T. (2006) 'Verification and the BWC: Last Gasp or Signs of Life?', *Arms Control Today* 36(7): 12–16.

Finnemore, M. and Sikkink, K. (1998) 'International Norm Dynamics and Political Change', *International Organization* 52(4): 887–917.

Finney, J.W. (1969a) 'Pentagon Denied Fund to Develop Chemical Arms', *New York Times*, 4 July: 1, 4.

Finney, J.W. (1969b) 'Bird Study Held Screen for Army', *New York Times*, 5 February: 42.

Finney, J.W. (1971) 'Opposition Seen to Gas-War Pact', *New York Times*, 17 March: 13.

Finney, J.W. (1973) 'U.S. Chemists Now Support Ban on Tear-Gas and Herbicide Use', *New York Times*, 8 October: 7.

Finney, J.W. (1974a) 'Fulbright Looks to Senate Vote on '25 Pact Banning Germ War', *New York Times*, 8 December: 12.

Finney, J.W. (1974b) 'Senate Committee Votes 1925 Chemical War Ban', *New York Times*, 13 December: 1, 2.

Foot, R. (1991) 'Making Known the Unknown War: Policy Analysis of the Korean Conflict in the Last Decade', *Diplomatic History* 15(3): 411–31.

Foot, R. (2005) 'Human Rights and Counterterrorism in Global Governance: Reputation and Resistance', *Global Governance* 11(3): 291–310.

Forrestal, J. (1949) 'Text of Forrestal Statement Minimizing the Immediate Peril of a Super "Germ" Weapon', *New York Times*, 13 March: 37.

Foster, W.C. (1966a) 'Memorandum from the Director of Defense Research and Engineering, Department of Defense (Foster) to the Assistant Secretary of Defense for International Security Affairs (McNaughton)', *Foreign Relations of the United States, 1964–1968, Volume X, National Security Policy*: https://history.state.gov/historicaldocuments/frus1964-68v10/d121.

Foster, W.C. (1966b) 'Statement by the ACDA Director Foster to the First Committee of the General Assembly: Use of Chemical and Biological Weapons', *Documents on Disarmament 1966*: 759–62.

Foster, W.C. (1969) 'Statement by Ambassador Foster', *Department of State Bulletin* LX(1543): 58–62.

Foucault, M. (1995) *Discipline and Punish: The Birth of the Prison*. Translated by Sheridan, A. New York: Vintage Books.

Foucault, M. (2003a) *Society Must Be Defended: Lectures at the College de France, 1975–76*. Translated by Macey, D. New York: Picador.

Foucault, M. (2003b) *The Birth of the Clinic: An Archaeology of Medical Perception*. Translated by Sheridan, A.M. London: Routledge.

Foucault, M. (2004) *The Birth of Biopolitics: Lectures at the College de France, 1978–79*. Translated by Burchell, G. Basingstoke: Palgrave Macmillan.

Fox, L.A. (1933) 'Bacterial Warfare: The Use of Biologic Agents in Warfare', *Military Surgeon* 72(3): 189–207.

Frank, F.R. (1975) *U.S. Arms Control Policymaking: The 1972 Biological Weapons Convention Case* [PhD Thesis, Stanford University].

Frankenberg, G. (2008) 'Torture and Taboo: An Essay Comparing Paradigms of Organized Cruelty', *The American Journal or Comparative Law* 56(2): 403–22.

Frazer, J.G. (1875) 'Taboo', *Encyclopaedia Britannica, 9th Edition*.

Frazer, J.G. (2013) *The Golden Bough: A Study of Magic and Religion*. London: Macmillan and Co.

Freedman, L. (2006) 'The Transformation of Strategic Affairs', *The Adelphi Papers* 45(379): 5–93.

Freedman, L. (2013) 'Disarmament and Other Nuclear Norms', *The Washington Quarterly* 36(2): 93–108.

French, F.C. (1907) 'A Factor in the Evolution of Morals', *The Journal of Philosophy, Psychology and Scientific Methods* 4(8): 203–9.
Freud, S. (2001a) 'Fetishism', in Strachey, J. (ed.), *The Standard Edition of the Complete Works of Sigmund Freud: Volume XXI (1927–1931)*. London: Vintage: 152–7.
Freud, S. (2001b) 'Three Essays of the Theory of Sexuality', in Strachey, J. (ed.), *The Standard Edition of the Complete Works of Sigmund Freud: Volume VII (1901–1905)*. London: Vintage: 135–246.
Freud, S. (2012) *Totem and Taboo: Resemblances Between the Psychic Lives of Savages and Neurotics*. London: George Routledge and Sons.
Fulbright, J.W. (1971a) 'Memorandum for the Secretary of State, the Secretary of Defense: 1925 Geneva Protocol': https://www.cia.gov/library/readingroom/docs/CIA-RDP73B00296R000400010034-0.pdf
Fulbright, J.W. (1971b) 'Letter to the President', *Congressional Record – Senate*, 27 July: 27437–8.
Fuller, H. (2016) 'Pandemic Cities: Biopolitical Effects of Changing Infection Control in post-SARS Hong Kong', *The Geographical Journal* 182(4): 342–52.
Furmanski, M. and Wheelis, M. (2009) 'Allegations of Biological Weapons Use', in Wheelis, M., Roizsa, L., and Dando, M. (eds), *Deadly Cultures: Biological Weapons since 1945*. Cambridge, MA: Harvard University Press: 252–83.
G2 (1941) 'G-2 to Office of the Surgeon General: Japanese Attempts to Secure Virulent Strains of Yellow Fever Virus', Record Group 112, *National Archives*.
Galamas, F. (2008) 'Biological Weapons, Nuclear Weapons and Deterrence: The Biotechnology Revolution', *Contemporary Strategy* 27(4): 315–23.
Gandhi, M. (2009) *Wisdom of Gandhi*. New Delhi: Ocean Books.
Gantz, W. (1978) 'The Movement of Taboos: A Message-Orientated Approach', Paper presented at the *Annual Meeting of the Association for Education in Journalism*, Seattle, Washington, 13–16August.
Geissler, E., van Courtland Moon, J.E., and Pearson, G.S. (1999) 'Lessons from the History of Biological and Toxin Warfare', in Geissler, E. and van Courtland Moon, J.E. (eds), *Biological and Toxin Weapons: Research, Development and Use from the Middle Ages to 1945*. Oxford: Oxford University Press, 255–76.
Gentili, A. (1933) *De iure belli libri tres*. Translated by Rolfe, J. Oxford: The Clarendon Press.
Gibler, D.M. (2008) 'The Costs of Reneging: Reputation and Alliance Formation', *The Journal of Conflict Resolution* 52(3): 426–54.
Gilman, D.C. (1901) 'Francis Lieber', *The Yale Law Journal* 10(7): 271–4.
Gizewski, P. (1996) 'From Winning Weapon to Destroyer of Worlds: The Nuclear Taboo in International Politics', *International Journal* 51(3): 397–419.
Gleason, S.E. (1956) 'Memorandum of Discussion at the 277th Meeting of the National Security Council, Washington, February 27, 1956', *Foreign Relations of the United States, 1955–57, National Security Policy, Volume XIX*: 201–18.
Gleason, S.E. (1958) 'Memorandum of Discussion at the 412th Meeting of the National Security Council', *Foreign Relations of the United States, 1958–1960, National Security Policy; Arms Control and Disarmament, Volume III*: 238–53.
Glennon, M.J. (2005) 'How International Rules Die', *The Georgetown Law Journal* 93: 939–91.
Goffman, E. (1990) *Stigma: Notes on the Management of Spoiled Identity*. London: Penguin.
Goldblat, J. (1997) 'The Biological Weapons Convention: An Overview', *International Review of the Red Cross* 37(318): 251–65.
Goldman, D.I. (2009) 'The Generals and the Germs: The Army Leadership's Response to Nixon's Review of Chemical and Biological Warfare', *The Journal of Military History* 73(2): 531–69.

Goldsmith, J.L. and Posner, E.A. (2002) 'Moral and Legal Rhetoric in International Relations: A Rational Choice Perspective', *The Journal of Legal Studies* 31(S1): S115–39.

Gordin, M.D. (1997) 'The Anthrax Solution: The Sverdlovsk Incident and the Resolution of a Biological Weapons Controversy', *Journal of the History of Biology* 30(3): 441–80.

Gorman, J. (2012) 'Survival's Ick Factor', *New York Times*, 23 January: https://www.nytimes.com/2012/01/24/science/disgusts-evolutionary-role-is-irresistible-to-researchers.html?_r=1&partner=rss&emc=rss&pagewanted=all

Gould, R. and Connell, N. (1997) 'The Public Health Effects of Biological Weapons', in Levy, B.S. and Sidel V.W. (eds), *War and Public Health*. Oxford: Oxford University Press, 98–119.

Graeber, D. (2005) 'Fetishism as Social Creativity: or, Fetishes are Gods in the Process of Construction', *Anthropological Theory* 5(4): 407–38.

Graham, T. (2002) 'Biological Weapons, War Crimes, and WWI', *Science* 296(5571): 1236–7.

Greco, M. (2019) 'On Illness and Value: Biopolitics, Psychosomatics, Participating Bodies', *Medical Humanities* 45 (2): 107–15.

Greenberg, D.S. (1968) 'Certain Styles of Killing', *New York Times*, 9 June: 6, 42.

Greenspan, P. (2015) 'Confabulating the Truth: In Defense of "Defensive" Moral Reasoning', *The Journal of Ethics* 19(2): 105–23.

Grob, G.N. (1977) 'The Social History of Medicine and Disease in America: Problems and Possibilities', *Journal of Social History* 10(4): 391–409.

Gross, E.A. (1952a) 'The Soviet Germ Warfare Campaign: The Strategy of the Big Lie', *Department of State Bulletin* XXVII(683): 153–9.

Gross, E.A. (1952b) 'U.S. Proposes Investigation of Bacteriological Warfare Charges', *Department of State Bulletin* XXVII(680): 32–7.

Gross, E.A. (1953) 'Disarmament as One of the Vital Conditions of Peace', *Department of State Bulletin* XXVIII(718): 476–9.

Grotius, H. (2005) *The Rights of War and Peace: Book I*. Indianapolis: Liberty Fund.

Grygiel, J.J. (2006) *Great Powers and Geopolitical Change*. Baltimore: Johns Hopkins University Press.

Guhin, M. (1969) 'Memorandum from Michael Guhin of the National Security Council Staff to the Assistant to the President for National Security Affairs (Kissinger)', *Foreign Relations of the United States, 1969–176, Volume E-2, Documents on Arms Control and Nonproliferation, 1969–1972*: https://history.state.gov/historicaldocuments/frus1969-76ve02/d170

Guhin, M. (1970) 'Memorandum for Dr. Kissinger: Dr. DuBridge's Comments re Reactions to the President's Chemical Warfare and Biological Research Announcement', *National Security Archive*: http://nsarchive2.gwu.edu.946elmp01.blackmesh.com/NSAEBB/NSAEBB58/RNCBW18.pdf

Guhin, M. (1971) 'Memorandum from Michael Guhin of the National Security Council Staff to the President's Assistant for National Security Affairs (Kissinger)', *Foreign Relations of the United States, 1969–176, Volume E-2, Documents on Arms Control and Nonproliferation, 1969–1972*: https://history.state.gov/historicaldocuments/frus1969-76ve02/d240

Guhin, M. (1972a) 'Memorandum From Michael Guhin of the National Security Council Staff to the President's Assistant for National Security Affairs (Kissinger)', *Foreign Relations of the United States, 1969–176, Volume E-2, Documents on Arms Control and Nonproliferation, 1969–1972*: https://history.state.gov/historicaldocuments/frus1969-76ve02/d251

Guhin, M. (1972b) 'Memorandum from Michael Guhin of the National Security Council Staff to the President's Assistant for National Security Affairs (Kissinger)', *Foreign Relations of the United States, 1969–176, Volume E-2, Documents on Arms Control and Nonproliferation, 1969–1972*: https://history.state.gov/historicaldocuments/frus1969-76ve02/d272

Guhin, M. (1974) 'Memorandum from Michael Guhin of the National Security Council Staff to the President's Deputy Assistant for National Security Affairs (Scowcroft)', *Foreign Relations of the United States, 1969–1976, Volume E-14, Part 2, Documents on Arms Control and Nonproliferation, 1973–1976*: https://history.state.gov/historicaldocuments/frus1969-76ve14p2/d46

Guillemin, J. (1999) *Anthrax: The Investigation of a Deadly Outbreak*. Berkeley: University of California Press.

Guillemin, J. (2005) *Biological Weapons: From the Invention of State-Sponsored Programs to Contemporary Bioterrorism*. New York: Columbia University Press.

Guillemin, J. (2017) *Hidden Atrocities: Japanese Germ Warfare and American Obstruction of Justice at the Tokyo Trial*. New York: Columbia University Press.

Gusinger, A. and Smith, A. (2002) 'Honest Threats: The Interaction of Reputation and Political Institutions in International Crises', *The Journal of Conflict Resolution* 46(2): 175–200.

Guyatt, D. (2005) 'Anti-Personnel "Soft-Kill" Em Weaponry', *Journal of International Issues* 9(4): 41–57.

Guzman, A.T. (2002) 'A Compliance-Based Theory of International Law', *California Law Review* 90(6): 1823–88.

Haas, P.H. (1998) 'Compliance with EU Directives: Insights from International Relations and Comparative Politics', *Journal of European Public Policy* 5(1): 17–37.

Hafner-Burton, E. and Montgomery, A.H. (2006) 'Power Positions: International Organizations, Social Networks, and Conflict', *The Journal of Conflict Resolution* 50(1): 3–27.

Haidt, J. et al. (1997) 'Body, Psyche, and Culture: The Relationship between Disgust and Morality', *Psychology and Developing Societies* 9(1): 107–31.

Haig, A.M. (1981) 'Address before the Berlin Press Association, West Berlin', *Department of State Bulletin* 81(2056): 42–9.

Hajar, R. (2017) 'The Physician's Oath: Historical Perspectives', *Heart Views* 18(4): 154–9.

Hajer, M. (1997) *The Politics of Environmental Discourse: Ecological Modernization and the Policy Process*. Oxford: Oxford University Press.

Haldane, J. (1925) *Callinicus: A Defence of Chemical Weapons*. London, Kegan Paul.

Haldeman, H.R. (1972) 'Conversation between President Nixon and the President's Chief of Staff (Haldeman)', *Foreign Relations of the United States, 1969–176, Volume E-2, Documents on Arms Control and Nonproliferation, 1969–1972*: https://history.state.gov/historicaldocuments/frus1969-76ve02/d254

Hall, T. and Ross, A.A.G. (2015) 'Affective Politics after 9/11', *International Organization* 69(4): 847–79.

Halliday, J. (1984) 'Anti-Communism and the Korean War, 1950–53', *The Socialist Register* 1984 21: 130–63.

Halperin, M. (1969) 'Memorandum from Morton Halperin of the National Security Council Staff to the President's Assistant for National Security Affairs (Kissinger)', *Foreign Relations of the United States, 1969–176, Volume E-2, Documents on Arms Control and Nonproliferation, 1969–1972*: https://history.state.gov/historicaldocuments/frus1969-76ve02/d145

Hamilton, T.J. (1952a) 'Germ-Warfare Canard Grows to Major Issue', *New York Times*, 13 July: E6.

Hamilton, T.J. (1952b) 'U.S. once Spurred Protocol It now Fights on Germ War', *New York Times*, 20 June: 3.

Hamilton, T.J. (1969) 'U.S. Restores Sentence to a Nixon Text on Arms', *New York Times*, 31 July: 8.

Hamlin, C. (2012) 'The Cholera Stigma and the Challenge of Interdisciplinary Epistemology: From Bengal to Haiti', *Science as Culture* 21(4): 445–74.

Han, Q. and Curtis, D.R. (2020) 'Social Responses to Epidemics Depicted by Cinema', *Emerging Infectious Diseases* 26(2): 389–94.

Hanrahan, J. (1969) 'Germs, War and Ft. Detrick', *The Washington Post*, 13 July3: 1, A20.

Hansen, L.M. (1990) 'Biological and Toxin Weapons: Arms Control, Stability, and Western Security', *Politics and the Life Sciences*, 9(1): 47–58.

Harbour, F.V. (1995) 'Islamic Principles and the Chemical Weapons Convention of 1993', *The Journal of Religious Ethics* 23(1): 69–92.

Harrington de Santana, A. (2009) 'Nuclear Weapons as the Currency of Power: Deconstructing the Fetishism of Force', *Nonproliferation Review* 16(3): 325–45.

Harris, E.D. (1987) 'Sverdlovsk and Yellow Rain: Two Cases of Soviet Noncompliance?', *International Security* 11(4): 41–95.

Harris, H.E. (1973) 'Modern Weapons and the Law of Land Warfare', *Military Law and Law of War Review* 12(1): 7–50.

Harris, R. and Paxman, J. (2002) *A Higher Form of Killing: The Secret History of Gas and Germ Warfare*. London: Arrow.

Harris, S. (2002) *Factories of Death: Japanese Biological Warfare, 1932–1945, and the American Cover-up*. London: Routledge.

Hart, J. and Trapp, R. (2012) 'Science, Technology, and the Biological Weapons Convention', *Arms Control Today* 42(8): 15–21.

Hashmi, S.H. and Lee, S.P. (2004) 'Introduction', in Hashmi, S.H. and Lee, S.P. (eds), *Ethics of Mass Destruction: Religious and Secular Perspectives*. Cambridge: Cambridge University Press, 1–15.

Hawthorne, N. (1850) *The Scarlett Letter*: https://ebooks.adelaide.edu.au/h/hawthorne/nathaniel/h39s/

Hay, A. (1999) 'Simulants, Stimulants and Diseases: The Evolution of the United States Biological Warfare Programme', *Medicine, Conflict and Survival* 15(3): 198–214.

Haynes, N. (2020) *Pandora's Jar: Women in the Greek Myths*. London: Picador.

Helman, G. (1980) 'Telegram from the Mission in Geneva to the Department of State', *Foreign Relations of the United States, 1977–1980, Volume XXVI, Arms Control and Nonproliferation*: https://history.state.gov/historicaldocuments/frus1977-80v26/d119

Herman, E.S. and Rutman, R.J. (1967) 'University of Pennsylvania's CB Warfare Controversy', *BioScience* 17(8): 526–9.

Hersh, S.M. (1968a) *Chemical and Biological Warfare: America's Secret Arsenal*. London: MacGibbon & Kee.

Hersh, S.M. (1968b) 'Chemical and biological weapons—The Secret Arsenal', *New York Times*, 25 August: 27, 82–7, 90, 92.

Hersh, S.M. (1969) 'Dare We Develop Biological Weapons?', *New York Times*, 8 September: 28–9, 78–83.

Hickerson, J.D. (1952) 'Memorandum by the Assistant Secretary of State for United Nations Affairs (Hickerson) to the Deputy Under Secretary of State (Matthews)', *Foreign Relations of the United States, 1952–1954, Korea, Volume XV, Part I*: 210–12.

Hillenkoetter, R.H. (1948) 'Memorandum Report for the Secretary of National Defense': https://www.cia.gov/library/readingroom/document/cia-rdp83-01034r000200080001-1

Hoffman, B. (1997) 'Terrorism and WMD: Some Preliminary Hypotheses', *Nonproliferation Review* 4(3): 45–53.

Holden, L. (2001) *Taboos: Structure and Rebellion*. London: The Institute for Cultural Research.

Homan, R. (1969) 'Panel Bars Funds for Germ Warfare', *The Washington Post*, 4 July: A1.

Hopf, T. (1998) 'The Promise of Constructivism in International Relations Theory', *International Security* 23(1): 171–200.

Horberg, E.J. et al. (2009) 'Disgust and the Moralization of Purity', *Journal of Personality and Social Psychology* 97(6): 963–76.

Horowitz, M.C. and Narang, N. (2014) 'Poor Man's Atomic Bomb? Exploring the Relationship between "Weapons of Mass Destruction"', *Journal of Conflict Resolution* 58(3): 509–35.

Hugh-Jones, M. (1992) 'Wickham Steed and German Biological Warfare Research', *Intelligence and National Security* 7(4): 379–402.

Hull, W.I. (1909) 'The Influence of Peace Power upon History', *The Advocate of Peace* 71(4): 84–7.

Hull, W.I. (1910) 'America and the Hague Conferences', *Maryland Peace Society Quarterly* 2: 3–16.

Hurrell, A. and Kingsbury, B. (1992) *The International Politics of the Environment*. Oxford: Oxford University Press.

Huth, P. and Russett (1984) 'What Makes Deterrence Work? Cases from 1900 to 1980', *World Politics* 36(4): 496–526.

Hymans, J. (2006) *The Psychology of Nuclear Proliferation: Identity, Emotions and Foreign Policy*. Cambridge: Cambridge University Press.

Ignatieff, G. (1974) 'Negotiating Arms Control', *International Journal* 30(1): 92–101.

Ikle, F.C. (1974) 'Statement by the Director of the Arms Control and Disarmament Agency (Ikle)', *Foreign Relations of the United States, 1969–1976, Volume E-14, Part 2, Documents on Arms Control and Nonproliferation, 1973–1976*: https://history.state.gov/historicaldocuments/frus1969-76ve14p2/d104

Ingram, A. (2009) 'The Geopolitics of Disease', *Geography Compass* 3(6): 2084–97.

Jacobson, J.K. (2008) 'Why do States Bother to Deceive? Managing Trust at Home and Abroad', *Review of International Studies* 34(2): 337–61.

Jay, T. (2009) 'The Utility and Ubiquity of Taboo Words', *Perspectives on Psychological Science* 4(2): 153–61.

Jefferson, C. (2009) *The Taboo of Chemical and Biological Weapons: Nature, Norms and International Law* [DPhil Thesis, University of Sussex].

Jeffrey, R. (2014) *Reason and Emotion in International Ethics*. Cambridge: Cambridge University Press.

Johnson, P.C. (2000) 'The Fetish and McGwire's Ball', *Journal of the American Academy of Religion* 68(2): 243–64.

Johnstone, L.C. (1971) 'Ecocide and the Geneva Protocol', *Foreign Affairs* 49(4): 711–20.

Jones, C.P. (1987) 'Stigma: Tattooing and Branding in Graeco-Roman Antiquity', *The Journal of Roman Studies* 77: 139–55.

Jones, D.P. (1980) 'American Chemists and the Geneva Protocol', *ISIS* 71(3): 426–40.

Jones, H. (2015) 'Violent Transgression and the First World War', *Studies: An Irish Quarterly Review* 104(414): 124–43.

Kaempffert, W. (1946) 'Bacterial Warfare Poses a Problem as Hard to Solve as That of the Atomic Bomb', *New York Times*, 12 January: 14.

Kaempffert, W. (1947) 'Deadly Germs Described', *New York Times*, 19 May: 11.

Kaempffert, W. (1952) 'Charges that the Allies Used Germ Warfare in Korea Constitute a Most Unlikely Story', *New York Times*, 16 March: E11.

Kang, D.C. (2004) 'The Theoretical Roots of Hierarchy in International Relations', *Australian Journal of International Affairs* 58(3): 337–52.

Kass, L. (1997) 'The Wisdom of Repugnance', *The New Republic*, 2 June: 17–26.

Kastenmeier, R.W. (1969) 'Increased Emphasis by Defense Department on CBR Program', *Congressional Record – House*, 24 April: 10271–3.

Katzenstein, P. (1996) 'Introduction: Alternative Perspectives on National Security', in Katzenstein, P. (ed.), *The Culture of National Security: Norms and Identity in World Politics*, New York: Columbia University Press, 1–32.

Kaufman, D.B. (1932) 'Poisons and Poisoning among the Romans', *Classical Philology* 27(2): 156–67.

Keane, W. (1997) 'From Fetishism to Sincerity: On Agency, the Speaking Subject, and their Historicity in the Context of Religious Conversion', *Comparative Studies in Society and History* 39(4): 674–93.

Kearney, A. (1992) 'Re-Evaluating the BWC: The Changing Threat of Biological Weapons', *Harvard International Review* 14(3): 50–3.

Keefer, S.A. (2006) 'Building the Palace of Peace: The Hague Conference of 1899 and Arms Control in the Progressive Era', *Journal of the History of International Law* 8(1): 1–18.

Keir, E. and Mercer, J. (1996) 'Setting Precedents in Anarchy: Military Intervention and Weapons of Mass Destruction', *International Security* 20(4): 77–106.

Kekes, J. (1992) 'Disgust and Moral Taboos', *Philosophy* 67(262): 431–46.

Kelle, A. (1997) 'Developing Control Regimes for Chemical and Biological Weapons', *The International Spectator* 32(3-4): 137–57.

Kelle, A. (2007) 'Securitization of International Public Health: Implications for Global Health Governances and the Biological Weapons Prohibition Regime', *Global Governance* 13(2): 217–35.

Kellman, B. (2007) *Bioviolence: Preventing Biological Terror and Crime*. Cambridge: Cambridge University Press.

Kellogg, F.B. (1929) 'The Secretary of State to the Ambassador in Belgium (Gibson)', *Papers Relating to the Foreign Relations of the United States, 1929, Volume 1*: 70–8.

Kelly, D. (2011) *Yuck! The Nature and Moral Significance of Disgust*. Cambridge, MA: MIT Press.

Kennan, G.F. (1952) 'The Ambassador in the Soviet Union (Kennan) to the Department of State', *Foreign Relations of the United States, 1952-1954, Eastern Europe; Soviet Union; Eastern Mediterranean, Volume VIII*: 971–7.

Kennedy, J.F.K. (1960) 'Remarks by Senator Kennedy at the University of New Hampshire', *Documents on Disarmament 1960*: 58–65.

Kennedy, J.F.K. (1963) 'Strength for Peace and Strength for War', *Department of State Bulletin* XLIX(1271): 694–7.

Kennedy, P.P. (1951) 'Book on Germ Warfare Gives Defense Tips', *New York Times*, 15 March: 31.

Kervers, O. (2003) 'Strengthening Compliance with the Biological Weapons Convention: The Draft Protocol', *Journal of Conflict & Security Law* 8(1): 161–200.

King, J. (1793) *A Voyage to the Pacific Ocean; Undertaken by Command of his Majesty for Making Discoveries in the Northern Hemisphere: Performed under the Direction of Captains Cook, Clerke, and Gore, in the Years 1776, 1777, 1778, 1779, 1780. Volume III*. London: Champante and Withrow.

Kissinger (1971) 'Memorandum from the President's Assistant for National Security Affairs (Kissinger) to President Nixon', *Foreign Relations of the United States, 1969–1976, Volume E-2, Documents on Arms Control and Nonproliferation, 1969–1972*: https://history.state.gov/historicaldocuments/frus1969-76ve02/d225

Kissinger, H. (1969a) 'Memorandum from the President's Assistant for National Security Affairs (Kissinger) to President Nixon', *Foreign Relations of the United States, 1969–1976, Volume E-2, Documents on Arms Control and Nonproliferation, 1969–1972*: https://history.state.gov/historicaldocuments/frus1969-76ve02/d140

Kissinger, H. (1969b) 'National Security Study Memorandum 59', *Foreign Relations of the United States, 1969–1976, Volume XXXIV, National Security Policy 1969–1972*: https://history.state.gov/historicaldocuments/frus1969-76ve02/d141.

Kissinger, H. (1969c) 'Notes of Telephone Conversation between the President's Assistant for National Security Affairs (Kissinger) and Secretary of Defense Laird', *Foreign Relations of the United States, 1969–1976, Volume E-2, Documents on Arms Control and Nonproliferation, 1969–1972*: https://history.state.gov/historicaldocuments/frus1969-76ve02/d160

Kissinger, H. (1969d) 'Memorandum from the President's Assistant for National Security Affairs (Kissinger) to President Nixon', *Foreign Relations of the United States, 1969–1976, Volume XXXIV, National Security Policy 1969–1972*: https://history.state.gov/historicaldocuments/frus1969-76v34/d127.

Kissinger, H. (1969e) 'Notes of Telephone Conversation Between the President's Assistant for National Security Affairs (Kissinger) and Secretary of Defense Laird', *Foreign Relations of the United States, 1969–1976, Volume E-2, Documents on Arms Control and Nonproliferation, 1969–1972*: https://history.state.gov/historicaldocuments/frus1969-76ve02/d171

Kissinger, H. (1969f) 'Talking Points Prepared for the President's Assistant for National Security Affairs (Kissinger)', *Foreign Relations of the United States, 1969–1976, Volume E-2, Documents on Arms Control and Nonproliferation, 1969–1972*: https://history.state.gov/historicaldocuments/frus1969-76ve02/d153

Kissinger, H. (1969g) 'Notes of Telephone Conversation between the President's Assistant for National Security Affairs (Kissinger) and Secretary of Defense Laird', *Foreign Relations of the United States, 1969–1976, Volume E-2, Documents on Arms Control and Nonproliferation, 1969–1972*: https://history.state.gov/historicaldocuments/frus1969-76ve02/d159

Kissinger, H. (1969h) 'Memorandum from the President's Assistant for National Security Affairs (Kissinger) to President Nixon', *Foreign Relations of the United States, 1969–1976, Volume E-2, Documents on Arms Control and Nonproliferation, 1969–1972*: https://history.state.gov/historicaldocuments/frus1969-76ve02/d63

Kissinger, H. (1970) 'Memorandum from the President's Assistant for National Security Affairs (Kissinger) to President Nixon', *Foreign Relations of the United States, 1969–1976, Volume E-2, Documents on Arms Control and Nonproliferation, 1969–1972*: https://history.state.gov/historicaldocuments/frus1969-76ve02/d200

Kleitman, N. (1950) 'The Big Lie', *Science* 112(2912): 475–6.

Klotz, L.C. and Sylvester, E.J. (2009) *Breeding Bio Insecurity: How U.S. Biodefense is Exporting Fear, Globalizing Risk, and Making Us All Less Secure*. Chicago: University of Chicago Press.

Knollenberg, B. (1954) 'General Amherst and Germ Warfare', *The Mississippi Valley Historical Review* 41(3): 489–94.

Koblentz, G. (2009) *Living Weapons: Biological Warfare and International Security*. Ithaca: Cornell University Press.

Koblentz, G. (2013) 'Regime Security: A New Theory for Understanding the Proliferation of Chemical and Biological Weapons', *Contemporary Security Policy* 34(3): 501–25.

Koch, G. (2003) 'Between Fear of Contact and Self-Preservation: Taboo and Its Relation to the Dead', *New German Critique* 90: 71–83.

Kolnai, A. (2004a) 'Disgust, in Korsmeyer, C. and Smith, B. *On Disgust*. Chicago: Open Court, 29–91.

Kolnai, A. (2004b) 'The Standard Modes of Aversion: Fear, Disgust, and Hatred', in Korsmeyer, C. and Smith, B. *On Disgust*. Chicago: Open Court, 93–110.

Kong, B. (2019) 'Pandemic as Method', *Prism* 16(2): 368–89.

Korsmeyer, C. and Smith, B (2004) 'Visceral Values: Aurel Kolnai on Disgust', in Korsmeyer, C. and Smith, B. *On Disgust*. Chicago: Open Court, 1–25.

Krasner, S. (1999) *Sovereignty: Organized Hypocrisy*. Princeton: Princeton University Press.

Kratochwil, F. (1995) *Rules, Norms, and Decisions*. Cambridge: Cambridge University Press.

Krickus, R.J. (1965) 'On the Morality of Chemical/Biological War', *The Journal of Conflict Resolution* 9(2): 200–10.

Kristeva, J. (1982) *Powers of Horror: An Essay on Abjection*. Translated by Roudiez, L.S. New York: Columbia University Press.

Kroll, D.J. (2006) *Securing Our Water Supply: Protecting a Vulnerable Resource*. Oklahoma: PenWell.

Laird, M. (1969a) 'Memorandum from Secretary of Defense Laird to the President's Assistant for National Security Affairs (Kissinger), *Foreign Relations of the United States, 1969–1976, Volume E-2, Documents on Arms Control and Nonproliferation, 1969–1972*: https://history.state.gov/historicaldocuments/frus1969-76ve02/d139

Laird, M. (1969b) 'Memorandum from Secretary of Defense Laird to the Assistant Secretary of Defense for International Security Affairs (Nutter)', *Foreign Relations of the United States, 1969–176, Volume E-2, Documents on Arms Control and Nonproliferation, 1969–1972*: https://history.state.gov/historicaldocuments/frus1969-76ve02/d143.

Lang, A. (1879) 'Mr. Max Mueller and Fetishism', *Mind* 4(16): 453–69.

Langer, E. (1967a) 'Chemical and Biological Warfare (II): The Weapons and the Policies', *Science* 155(3760): 299–303.

Langer, E. (1967b) 'Chemical and Biological Weapons: Once over Lightly on Capitol Hill', *Science* 156(3778): 1073–5.

Langer, E. (1968) 'United States', in Rose, E. (ed.), *CBW Chemical and Biological Warfare*. London: George B Harrap and Co Ltd, 119–29.

Lantis, J. and Wunderlich, C. (2018) 'Resiliency Dynamics of Norm Clusters: Norm Contestation and International Cooperation', *Review of International Studies* 44 (3): 570–93.

Laurence, W.L. (1947) 'Scientists Assail Bacteria Warfare', *New York Times*, 27 July: 12.

Lawrence, T.J. (1908) *International Problems and Hague Conferences*. London: J.M. Dent & Co.

Leach, E. (1964) 'Anthropological Aspects of Language: Animal Categories and Verbal Abuse', in Lenneberg, E.H. (ed.), *New Directions in the Study of Language*. Cambridge MA: MIT Press, 23–63.

Leahy, W.D. (1950) *I Was There*. New York: McGraw Hill.

Lederberg, J. (2001) 'Biological Warfare', *Emerging Infectious Diseases* 7(6): 1071–2.

Legro, J.W. (1997) 'Which Norms Matter? Revisiting the "Failure" of Internationalism', *International Organisation* 51(1): 31–63.

Leiker, J.J., Taub, D.E., and Gast, J. (1995) 'The Stigma of AIDS: Persons with AIDS and Social Distance', *Deviant Behavior* 16(4): 333–51.

Leitenberg, M. (1997) 'Biological Weapons, International Sanctions and Proliferation', *Asian Perspective* 21(3): 7–39.

Leitenberg, M. (1998) 'New Russian Evidence on the Korean War Biological Warfare Allegations: Background and Analysis', *Cold War International History Bulletin* 11: 185–99.

Leitenberg, M. (2000) 'The Korean War Biological Weapons Allegations: Additional Information and Disclosures', *Asian Perspective* 24(3): 159–72.

Leitenberg, M. (2002) 'Biological Weapons and Bioterrorism in the First Years of the Twenty-First Century', *Politics and the Life Sciences* 21(2): 3–27.

Leitenberg, M. (2008) 'False Allegations of U.S. Biological Weapons Use during the Korean War', in Clunan, A.S., Lavoy, P., and Martin, S. (eds), *Terrorism, War, or Disease? Unraveling the Use of Biological Weapons*. Stanford: Stanford University Press, 120–43.

Leitenberg, M., Zilinskas, R.A., and Kuhn, J.H. (2012) *The Soviet Biological Weapons Program: A History*. Cambridge, MA: Harvard University Press.

Leonard, J. (1986) 'Reviewing the Biological Weapons Convention', *Issues in Science and Technology* 3(1): 12–13.

Leonard, J.F. (2001) 'The BWC After the Protocol: Previewing the Review Conference: An ACA Press Conference', *Arms Control Today*, 31(10): 13–14.

LeVeck, B.L. and Narang, N. (2017) 'How International Reputation Matters: Revisiting Alliance Violations in Context', *International Interactions* 43(5): 797–821.

Levy, J.S. (1997) 'Prospect Theory, Rational Choice, and International Relations', *International Studies Quarterly* 41(1): 87–112.

Lewis, D.R. (2016) 'Biotechnology: An Era of Hopes and Fears', *Strategic Studies Quarterly* 10(3): 23–46.

Lewis, M. (1995) *Shame: The Exposed Self*. New York: Free Press.

Link, B.G. and Phelan, J.C. (2001) 'Conceptualizing Stigma', *Annual Review of Sociology* 27: 363–85.

Littlewood, J. (2005) *The Biological Weapons Convention: A Failed Revolution*. Aldershot: Ashgate.

Littlewood, J. (2007a) 'Biological Weapons: Much Ado and Little Action', *Minerva* 45(2): 191–203.

Littlewood, J. (2007b) 'Out of the Valley: Advancing the Biological Weapons Convention after the 2006 Review Conference', *Arms Control Today* 37(2): 12–16.

Lockwood, J.A. (2009) *Six Legged Soldiers: Using Insects as Weapons of War*. Oxford: Oxford University Press.

Lorenzini, D. (2021) 'Biopolitics in the Time of Coronavirus', *Critical Inquiry* 47(S2): S40–S45.

Lovell, S.P. (1944) 'Memorandum to: Colonel G Edward Buxton; Subject: OSS Biological Warfare Position', in CIA 'OSS—casualties from Allied phosphorous bombing of Berlin and Hamburg; potential for chemical and biological warfare; activities of censorship office; China assessments; 1942–95': https://www.cia.gov/library/readingroom/document/cia-rdp13x00001r000100430005-2

Lowitz, D.S. (1986) 'Review Conference Held on Biological and Toxin Weapons Convention', *Department of State Bulletin* 86(2117): 40–1.

Lubin, I. (1952) 'A Review of ECOSOC's 14th Session', *Department of State Bulletin* XXVII(687): 288–90.

Lusinchi, V. (1971) 'Russian Move Opens Way for Biological Arms Ban', *New York Times*, 31 March: 1, 16.

McCarthy, R.D. (1970) *The Ultimate Folly: War by Pestilence, Asphyxiation, and Defoliation*. London: Victor Gollancz.

McCauley, P.M. and Payne, R.A. (2010) 'The Illogic of the Biological Weapons Taboo', *Strategic Studies Quarterly* 4(1): 6–35.

MacGregor, G. (1953) 'Reds' Germ Propaganda Backfired When G.I. Captive Ate the Evidence', *New York Times*, 7 August: 1–2.

Machle, W. (1949) 'Memorandum from the Assistant Director for Scientific Intelligence (Machle) to Director of Central Intelligence Hillenkoetter', *Foreign Relations of the United States, 1945–50, Emergence of the Intelligence Establishment*: https://history.state.gov/historicaldocuments/frus1945-50Intel/d399

Mackby, J. (2018) 'Experts Debate Biological Weapons Challenges', *Arms Control Today* 48(7): 35–6.

MacLeod, R. (2017) 'The Genie and the Bottle: Reflections on the Fate of the Geneva Protocol in the United States, 1918–1928', in Friedrich, B., et al. (eds), *One Hundred Years of Chemical Warfare: Research, Deployment, Consequences*. Cham: Springer, 189–212.

Mancini, M.J. (2011) 'Francis Lieber, Slavery, and the "Genesis" of the Laws of War', *The Journal of Southern History* 77(2): 325–48.

Mangold, T. and Goldberg, J. (1999) *Plague Wars: The Terrifying Reality of Biological Warfare*. New York: St. Martin's Griffin.

Manisty, H.F. (1923) 'The Use of Poison Gas in War', *Transactions of the Grotius Society* 9: 17–28.
Mann, T. (1999) *The Magic Mountain*. London: Vintage Books.
Manzo, J.F. (2004) 'On the Sociology and Social Organization of Stigma: Some Ethnomethodological Insights', *Human Studies* 27(4): 401–16.
Marett, R.R. (1914) *The Threshold of Religion*. London: Methuen & Co.
Marsh, J.G. and Olsen, J.P (1998) 'The Institutional Dynamics of International Political Orders', *International Organization* 52(4): 943–69.
Marshall, D.A. (2010) 'Temptation, Tradition, and Taboo: A Theory of Sacralization', *Sociological Theory* 28(1): 64–90.
Martin, S. (2004) 'Weapons of Mass Destruction: A Brief Overview', in Hashmi, S. and Lee, S. (eds), *Ethics of Mass Destruction: Religious and Secular Perspectives*. Cambridge: Cambridge University Press, 16–42.
Marx, K. (1909) *Capital: A Critique of Political Economy. Volume I: The Process of Capitalist Production*. Chicago: Charles H. Kerr & Company.
Masuzawa, T. (2000) 'Troubles with Materiality: The Ghost of Fetishism in the Nineteenth Century', *Comparative Studies in Society and History* 42(2): 242–67.
Mauroni, A.J. (2000) *America's Struggle with Chemical-Biological Warfare*. Westport, CT: Praeger.
Valerius Maximus (1726) *Libri Novem Factorum Dictorumque*. Leiden: Samuel Luchtmans.
Mayo, C.W. (1953) 'The Role of Forced Confessions in the Communist "Germ Warfare" Propaganda Campaign', *Department of State Bulletin* XXIX(750): 641–7.
Mayor, A. (2009) *Greek Fire, Poison Arrows, and Scorpion Bombs: Biological and Chemical Warfare in the Ancient World*. New York: Overlook.
Mead, M. (1937) 'Tabu.' In Seligman, E.R.A. and Johnson, A. (eds), *Encyclopaedia of Social Sciences, Volume 14*. London: Macmillan, 502–5.
Meier, O. (2002) 'Verification of the Biological Weapons Convention: What Is Needed?', *Medicine, Conflict and Survival* 18(2): 175–93.
Menninghaus, W. (2003) *Disgust: Theory and History of a Strong Sensation*. New York: State University of New York Press.
Mercer, J. (1996) *Reputation and International Politics*. Ithaca, NY: Cornell University Press.
Mercer, J. (2014) 'Feeling Like a State: Social Emotion and Identity', *International Theory* 6(3): 515–35.
Merck, G.W. (1946) 'Biological Warfare: Report to the Secretary of War by Mr. George Merck, Special Consultant for Biological Warfare', *The Military Surgeon* 98(3): 237–42.
Merriam, E. (2014) 'The International Legal Regime Affecting Bioterrorism', *National Security Law Journal* 3(1): 1–46.
Meselson, M. et al. (1994) 'The Sverdlovsk Anthrax Outbreak of 1979', *Science* 266(5188): 1202–8.
Meselson, M. (1999) 'The Problem of Biological Weapons', *Bulletin of the American Academy of Arts and Sciences* 52(5): 46–58.
Meselson, M.S. and Robinson, J.P. (2008) 'The Yellow Rain Affair: Lessons from a Discredited Allegation', in Clunan, AS., Lavoy, P., and Martin, S. (eds), *Terrorism, War, or Disease? Unraveling the Use of Biological Weapons*. Stanford, CA: Stanford University Press, 72–96.
Meyer, P. (2018) 'Our WMD Treaties are Working', *Bulletin of the Atomic Scientists* 74(1): 27–31.
Meyers, J.S. (2015) 'Reputation Matters: Evidence from the Korean War', *Journal of International and Area Studies* 22(2): 19–37.
Miller, G.D. (2003) 'Hypotheses on Reputation: Alliance Choices and the Shadow of the Past', *Security Studies* 12(3): 40–78.

Miller, S.B. (2004) *Disgust: The Gatekeeper Emotion*. London: The Analytic Press.

Miller, W.I. (1997) *The Anatomy of Disgust*. Cambridge, MA: Harvard University Press.

Millet, P. (2003) 'The Biological Weapons Convention after November 2002', *Medicine, Conflict and Survival* 19(4): 303–17.

Millet, P. (2010) 'The Biological Weapons Convention: Securing Biology in the Twenty-First Century', *Journal of Conflict & Security Law* 15(1): 25–43.

Miskimmon, A., O'Loughlin, B., and Roselle, L. (2013) *Strategic Narratives: Communication Power and the New World Order*. London: Routledge.

Modini, E. (2005) 'Biowarfare as a Biopolitical Icon', *Poiesis Prax* 3(4): 242–55.

Monath, T.P. and Gordon, L.K. (1998) 'Strengthening the Biological Weapons Convention', *Science* 282(5393): 1423.

Montgomery, A.H. and Carpenter, C. (2021) 'Whence the Nuclear Taboo?', *International Studies Review* 23(3): 1089–92.

Moodie, M. (2001) 'The Soviet Union, Russia, and the Biological and Toxin Weapons Convention', *Nonproliferation Review* 8(1): 59–69.

Moore, J.N. (1972) 'Ratification of the Geneva Protocol on Gas and Bacteriological Warfare: Legal and Political Analysis', *Virginia Law Review* 58(3): 419–509.

Moreas, F. (1952) 'U.S. Is Enemy No. 1 in Red China, Indian Editor Finds on Tour There', *New York Times*, 14 June: 3.

Morris, R. (1977) *Uncertain Greatness: Henry Kissinger and American Foreign Policy*. New York: Harper & Row.

Mueller, H. (2021) 'Taboo or Tradition or What? A Critical Look at the Terminology and Conceptualization of Nuclear Nonuse', *International Studies Review* 23(3): 1082–5.

Mueller, J. (2001) *Retreat from Doomsday: The Obsolescence of Major War*. New York: Basic Books.

Mulley, F. (1968) 'Statement by the British Disarmament Minister (Mulley) to the Eighteen Nation Disarmament Committee [Extract], July 16, 1968', *Documents on Disarmament 1968*: 533–8.

Murphy, R.D. (1954) 'The Interrelationship of Military Power and Foreign Policy', *Department of State Bulletin* XXXI(792): 291–4.

Murrow, E.R. (1962) 'Memorandum from the Director of the United States Information Agency (Murrow) to the President's Special Assistant for National Security Affairs (Bundy)', *Foreign Relations of the United States, 1917–1972, Volume VI, Public Diplomacy 1961–1963*: https://history.state.gov/historicaldocuments/frus1917-72PubDipv06/d91

Nabi, R.L. (2002) 'The Theoretical Versus the Lay Meaning of Disgust: Implications for Emotion Research', *Cognition and Emotion* 16(5): 695–703.

Nabrit, J. (1966) 'Statement by the United States Representative (Nabrit) to the General Assembly: Use of Chemical and Biological Weapons', *Documents on Disarmament 1966*: 800–2.

NAS (1942a) 'Report of the WBC Committee; February 1942', Committees on Biological Warfare, 1941–48, Series 1 (Box 2), *National Academy of Sciences*.

NAS (1942b) 'Report of the WBC Committee; June 1942', Committees on Biological Warfare, 1941–48, Series 1 (Box 2), *National Academy of Sciences*.

Nassau, R.H. (1904) 'The Philosophy of Fetishism', *Journal of the Royal African Society* 3(11): 257–70.

Naughton, J.M. (1969) 'Nixon Renounces Germ Weapons, Orders Destruction of Stocks', *New York Times*, 26 November: 1, 3.

Naughton, J.M. (1970) 'Nixon Widens Ban on Germ Warfare to Include Toxins', *New York Times*, 15 February: 1, 23.

Neilands, J.B. (1970) 'Vietnam: Progress of the Chemical War', *Asian Survey* 10(3): 209–29.

Neilson, P.N. (1961) 'Memorandum from the Assistant Director, Far East (Neilson), to the Director of the United States Information Agency (Murrow)', *Foreign Relations of the United States, 1961–63, Volume I, Vietnam, 1961*: https://history.state.gov/historicaldocuments/frus1961-63v01/d265

Neinast, W.H. (1964) 'United States Use of Biological Warfare', *Military Law Review* 24: 1–46.

Nelson, G. (1969) 'Chemical and Biological Warfare', *Congressional Record – Senate*, 8 August: 23071–74.

Nichols, R.W. (2002) 'Allaying the Threat of Biological Weapons', *Science* 295(5559): 1464.

Nietzsche, F. (1910) *Human All-Too-Human: A Book for Free Spirits, Part 1*. London: T.N. Foulis.

Nixon, R. (1969) 'Chemical and Biological Defense Policies and Programs', *Department of State Bulletin* LXI(1590): 541.

Nixon, R. (1971a) 'National Center for Toxicological Research', *Weekly Compilation of Presidential Documents* 7(5): 123

Nixon, R. (1971b) 'Remarks Announcing the Conversion of Fort Detrick, Maryland, to a Center for Cancer Research', *Public Papers of the President of the United States 1971*: 1050–3.

Nixon, R. (1972a) 'Remarks by President Nixon on the Signing of the Biological Weapons Convention', *Documents on Disarmament 1972*: 138–9.

Nixon, R. (1972b) 'Arms Control Report Transmitted to the Congress', *Department of State Bulletin* LXVI(1708): 455.

Nolan, J. (2008) 'The Nuclear Taboo: The United States and the Non-Use of Nuclear Weapons Since 1945', *The Nonproliferation Review* 15(3): 539–42.

NSC (1950) 'NSC 62: Chemical Warfare Policy', Record Group 273, *National Archives*.

NSC (1953) 'NSC 147: Analysis of Possible Courses of Action in Korea', *Foreign Relations of the United States, 1952–1954, Korea, Volume XV, Part I*: https://history.state.gov/historicaldocuments/frus1952-54v15p1/d437

NSC (1956) 'National Security Council Report', *Foreign Relations of the United States, 1955–57, National Security Policy, Volume XIX*: https://history.state.gov/historicaldocuments/frus1955-57v19/d66

Nussbaum, M.C. (2006) *Hiding from Humanity: Disgust, Shame, and the Law*. Princeton, NJ: Princeton University Press.

Nys, E. (1911) 'Francis Lieber—His Life and His Work', *The American Journal of International Law* 5(2): 355–93.

O'Brien, W.V. (1962) 'Biological/Chemical Warfare and the International Law of War', *The Georgetown Law Journal* 51(1): 1–63.

O'Doherty, E.F. (1960) 'Taboo, Ritual and Religion', *An Irish Quarterly Review* 49(194): 131–43.

O'Tuathail, G. (2002) 'Theorizing Practical Geopolitical Reasoning: The Case of the United States' Response to the War in Bosnia', *Political Geography* 21(5): 601–28.

Oaten, M., Stevenson, R.J., and Case, T.I. (2009) 'Disgust as a Disease-Avoidance Mechanism', *Psychological Bulletin* 135(2): 303–21.

Oldendorf, W.H. (1962) 'On the Acceptability of a Device as a Weapon', *Bulletin of Atomic Scientists* 18(1): 35–7.

Oliver, K. (2003) 'Atrocity, Authenticity and American Exceptionalism: (Ir)rationalizing the Massacre at My Lai', *Journal of American Studies* 37(2): 247–68.

Olsthoorn, P. (2008) 'Honour, Face and Reputation in Political Theory', *European Journal of Political Theory* 7(4): 472–91.

ORO (1959) *Staff Paper ORO-SP 86: Attitudes and Policies on Chemical Warfare*.

Ovid (2004) *Fasti*. Translated by Boyle, A.J. and Woodard, R.D. London: Penguin.

Ovid (2014) *Metamorphoses*. Translated by Raeburn, D. London: Penguin.

Panke, D. (2010) 'Why Big States Cannot Do What They Want: International Legalization at Work', *International Politics* 47(2): 186–209.

Panke, D. and Petersohn, U. (2012) 'Why International Norms Disappear Sometimes', *European Journal of International Relations* 18(4): 719–42.

Panksepp, J. (2007) 'Criteria for Basic Emotions: Is DISGUST a Primary "Emotion"?', *Cognition and Emotion* 21(8): 1819–28.

Papagrigorakis, M.J. et al. (2006) 'DNA Examination of Ancient Dental Pulp Incriminates Typhoid Fever as a Probable Cause of the Plague of Athens', *International Journal of Infectious Disease* 10(3): 206–14.

Parachini, J. (2001) 'Non-Proliferation Policy and the War on Terrorism', *Arms Control Today*: https://www.armscontrol.org/act/2001-10/features/non-proliferation-policy-war-terrorism

Parachini, J. (2018) 'North Korea's CBW Program: How to Contend with Imperfectly Understood Capabilities', *PRISM* 7(3): 90–101.

Parrott, L. (1952a) 'Reds Accuse U.N. of Germ War Tests on Korea Captives', *New York Times*, 23 March: 1, 3.

Parrott, L. (1952b) 'Red China Accuses U.S. of War Crimes', *New York Times*, 13 March: 1, 3.

Paul, T.V. (2009) *The Tradition of Non-Use of Nuclear Weapons*. Stanford, CA: Stanford University Press.

Paul, T.V. (2010) 'Taboo or Tradition? The Non-Use of Nuclear Weapons in World Politics', *Review of International Studies* 36(4): 853–63.

Pels, P.J. (1998) 'The Spirit of Matter: On Fetish, Rarity, Fact and Fancy', in Spyer, P. (ed.), *Border Fetishisms: Material Objects in Unstable Spaces*. London: Routledge, 91–120.

Pencheva, D. and Maronitis, K. (2018) 'Fetishizing Sovereignty in the Remain and Leave Campaigns', *European Politics and Society* 19(5): 526–39.

Pescosolido, B.A. and Martin, J.K. (2015) 'The Stigma Complex', *Annual Review of Sociology* 41: 87–116.

Petrof, S. (2015) '"The Dialectics of Media Representation." *Je Suis Charlie* as Fetishization of an Image', *ESSACHESS Journal of Communication Studies* 8(2): 207–25.

Pharr, C. (1932) 'The Interdiction of Magic in Roman Law', *Transactions and Proceedings of the American Philological Association* 63: 269–95.

Piel, G. (1946) 'BW', *Life*, 18 November: 118–30.

Pietz, W. (1985) 'The Problem of Fetish, I', *RES* 9: 5–17.

Pitkin, H.F. (1987) 'Rethinking Reification', *Theory and Society* 16(2): 263–93.

Pivetti, M., Camodeca, M., and Rapino, M. (2016) 'Shame, Guilty, and Anger: Their Cognitive, Physiological, and Behavioral Correlates', *Current Psychology* 35(4): 690–9.

Plakias, A. (2013) 'The Good and the Gross', *Ethical Theory and Moral Practice* 16(2): 261–78.

Plato (1987) *The Republic*. Translated by Lee, D. London: Penguin.

Pool, R. (1990) 'Fetishism Deconstructed', *Etnofoor* 3(1): 114–27.

Powell, T. (2017) 'Biological Warfare in the Korean War: Allegations and Cover-up', *Socialism and Democracy* 31(1): 23–42.

Powell, T. (2019) 'Biological Warfare in Korea: A Review of the Literature', *Socialism and Democracy* 33(2): 67–107.

Press, D. (2007) *Calculating Credibility: How Leaders Assess Military Threats*. Ithaca, NY: Cornell University Press.

Press, D.G., Sagan, S.D., and Valentino, B.A. (2013) 'Atomic Aversion: Experimental Evidence on Taboos, Traditions, and the Non-Use of Nuclear Weapons', *The American Political Science Review* 107(1): 188–206.

Press, N. (1985) 'Haber's Choice, Hobson's Choice, and Biological Warfare', *Perspectives in Biology and Medicine* 29(1): 92–108.

Price, R. (1997) *The Chemical Weapons Taboo*. Ithaca, NY: Cornell University Press.

Primack, J. and von Hippel, F. (1974) *Advice and Dissent: Scientists in the Political Arena*. New York: Basic Books Inc.

PSB (1953) 'Memorandum of Conversation: State Department Exploitation of Communist BW Charges': https://www.cia.gov/library/readingroom/docs/CIA-RDP80R01731R003300190004-6.pdf

Quarles, D.A. (1954) 'Memorandum for Director, Central Intelligence Agency: Information on Soviet BW Activities': https://www.cia.gov/library/readingroom/docs/CIA-RDP80R01731R001300230026-9.pdf

Radcliffe-Brown, A.R. (1939) *Taboo*. Cambridge: Cambridge University Press.

Randolph, N. (2021) 'Making Live and Letting Die in COVID-19: The Biopolitics of Race, Space and Freedom', *SPECTRA* 8(2): 7–18.

Ranlet, P. (2000) 'The British, the Indians, and Smallpox: What Actually Happened at Fort Pitt in 1973?', *Pennsylvania History* 67(3): 427–41.

Raphael, A. (1969a) 'US drops a germ war plan', *The Guardian*, 13 November: 2.

Raphael, A. (1969b) 'US considering total ban on germ warfare', *The Guardian*, 29 September: 3.

Ray, S. (2021) 'The Virality of Pandemics: Reassembling the Social in the Anthropocene', *Society and Culture in South Asia* 7(1): 16–31.

Raymond, G.A. (1997) 'Problems and Prospects in the Study of International Norms', *Menshon International Studies Review* 41(2): 205–45.

Raymond, J. (1959) 'Pentagon Spurs Chemical Arms', *New York Times*, 9 August: 3.

Raymond, J. (1960) 'Pentagon Widens "Germ War" Study', *New York Times*, 2 November: 1, 3.

Raymond, J. (1965) 'Gas Weapon: Pro and Con', *New York Times*, 28 March: E4.

Reagan, R. (1982) 'An Agenda for Peace', *Department of State Bulletin* 82(2064): 39–42.

Reagan, R. (1983a) 'Interview for *Bunte Illustrierte* Magazine', *Department of State Bulletin* 83(2076): 23–6.

Reagan, R. (1983b) 'Progress in the Quest for Peace and Deterrence', *Department of State Bulletin* 83(2072): 28–32.

Reagan, R. (1983c) 'Visit of Sultan of Oman: Arrival Ceremony', *Department of State Bulletin* 83(2075): 65–6.

Reagan, R. (1983d) 'Interview of President Reagan: Intermediate-Range Nuclear Force Negotiations [Extract], May 26, 1983', *Documents on Disarmament 1983*: 444–7.

Reed, A. (2013) 'Afterword', *Social Analysis* 57(3): 150–5.

Reed, R. (1969) 'Army Admits its Nerve Gas Killed 6,000 Sheep', *New York Times*, 22 May: 14.

Reed, R. (1974) 'Bumpers Beats Fulbright in Arkansas; Faubus Trailing in Governor's Primary', *New York Times*, 29 May: 85.

Regis, E. (1999) *The Biology of Doom: The History of America's Secret Germ Warfare Project*. New York: Henry Holt and Company.

Reston, J. (1965) 'Washington: Just a Little Old "Benevolent Incapacitator"', *New York Times*, 24 March: 42.

Reston, J. (1969) 'The Pentagon: Nerve Gas in Okinawa', *New York Times*, 20 July: E12.

Reus-Smit, C. (2014) 'Emotions and the Social', *International Theory* 6(3): 568–74.

Revill, J. (2017) *Compliance Revisited: An Incremental Approach to Compliance in the Biological Toxin Weapons Convention, Occasional Paper No. 31*. James Martin Center for Nonproliferation Studies.

Revill, J. (2018) '"Muddling Though" in the Biological and Toxin Weapons Convention', *International* Politics 55(3–4): 386–401.

Ridgway, M.B. (1952) 'A Report on the Far East', *Department of State Bulletin* XXVI(676): 924–7.

Rimmington, A. (2000) 'Invisible Weapons of Mass Destruction: The Soviet Union's BW Programme and its Implications for Contemporary Arms Control', *The Journal of Slavic Military Studies* 13(3): 1–46.

Risse, T. and Wiener, A. (1999) '"Something Rotten" and the Social Construction of Social Constructivism: A Comment on Comments', *Journal of European Public Policy* 6(5): 775–82.

Rives, J.B. (2003) 'Magic in Roman Law: The Reconstruction of a Crime', *Classical Antiquity* 22(2): 313–39.

Roberts, B. (1992) 'Implementing the Biological Weapons Convention: Looking Beyond the Verification Issue', in Thraenert, O. (ed.), *The Verification of the Biological Weapons Convention: Problems and Perspectives*. Bonn: Friedrich Ebert Foundation, 101–6.

Robertson-Smith, W. (1884) *Lectures on the Religion of the Semites*. London: Adam & Charles Black.

Robinson, J., Guillemin, J., and Meselson, M. (1987) 'Yellow Rain: The Story Collapses', *Foreign Policy* 68: 100–17.

Robinson, J.P.P. (1981) 'Chemical Arms Control and the Assimilation of Chemical Weapons', *International Journal* 36(3): 515–34.

Roffey, R., Tegnell, A., and Elgh, F. (2002) 'Biological Warfare in Historical Perspective' *Clinical Microbiology and Infection* 8(8): 450–4.

Rogaski, R. (2002) 'Nature, Annihilation, and Modernity: China's Korean War Germ-Warfare Experience Reconsidered', *The Journal of Asian Studies* 61(2): 381–415.

Rogers, W.P. (1969) 'Telegram 168999 from the Department of State to the Mission to the United Nations', *Foreign Relations of the United States, 1969–176, Volume E-2, Documents on Arms Control and Nonproliferation, 1969–1972*: https://history.state.gov/historicaldocuments/frus1969-76ve02/d149

Rogers, W.P. (1970) 'Telegram 129723 from the Department of State to the Embassy in the Soviet Union', *Foreign Relations of the United States, 1969–176, Volume E-2, Documents on Arms Control and Nonproliferation, 1969–1972*: https://history.state.gov/historicaldocuments/frus1969-76ve02/d204

Rogers, W.P. (1971) 'Memorandum from Secretary of State Rogers to President Nixon', *Foreign Relations of the United States, 1969–176, Volume E-2, Documents on Arms Control and Nonproliferation, 1969–1972*: https://history.state.gov/historicaldocuments/frus1969-76ve02/d247

Roosevelt, F.D. (1939) 'The President Presents Plan No. 1 to Carry Out the Provisions of the Reorganization Act. April 25, 1939', *Public Papers of the Presidents of the United States 1939*: 245–74.

Roosevelt, F.D. (1943) 'Use of Poison Gas', *Department of State Bulletin* VIII(207): 507.

Roosevelt, F.D. (1947) 'Letter to the Senate on Chemicals in Warfare', *The American Presidency Project*: https://www.presidency.ucsb.edu/documents/letter-the-senate-chemicals-warfare

Root, E. (1913) 'Francis Lieber', *The American Journal of International Law* 7(3): 453–69.

Rose, S, (1972) 'The Real Significance of CBW', *Instant Research on Peace and Violence* 2(1): 9–16.

Rosebury, T. (1949) *Peace and Pestilence: Biological Warfare and How to Avoid It*. New York: McGraw-Hill.

Rosebury, T. (1963) 'Medical Ethics and Biological Warfare', *Perspectives in Biology and Medicine* 6(4): 512–23.

Rosebury, T. (1968) 'Chemical Weapons', *New York Times*, 28 April: E19.

Rosenbaum, D.E. (1969) 'Activist Germ War Foe: Matthew Stanley Meselson', *New York Times*, 26 November 17.

Rosenberg, C. (1986) 'Disease and Social Order in American: Perceptions and Expectations', *The Milbank Quarterly* 64(S1): 34–55.

Rosenthal, A.M. (1952) 'U.N. Silences Soviet on Germ Warfare', *New York Times*, 29 March: 1, 2.
Rosenthal, A.M. (1953) 'Germ War Inquiry Demanded of Red', *New York Times*, 28 March: 3.
Ross, A. (2014) *Mixed Emotions: Beyond Fear and Hatred in International Conflict*. Chicago: University of Chicago Press.
Rostow, E.V. (1981) 'Prospects for Arms Control', *Department of State Bulletin* 81(2057): 68–72.
Rostow. E.V. (1982) 'Arms Control in Proper Perspective', *Department of State Bulletin* 82(2062): 39–45.
Rothschild, J.H. (1964) *Tomorrow's Weapons: Chemical and Biological*. New York: McGraw-Hill.
Rothschild, J.H. (1968) 'More Discussion of Chemical Weapons Use', *New York Times*, 30 April: 32.
Rozin, P. (1997) 'Food for Thought: Paul Rozin's Research and Teaching at Penn State', *Penn State Arts & Sciences*, Fall: https://www.sas.upenn.edu/sasalum/newsltr/fall97/rozin.html
Rozin, P. and Fallon, A.E. (1987) 'A Perspective on Disgust', *Psychological Review* 94(1): 23–41.
Rozin, P., Haidt, J., and McCauley, C.R. (2008) 'Disgust', in Lewis, M., Haviland-Jones, J.M., and Barrett, L.F. (eds), *Handbook of Emotions*, 3rd Edition. New York: Guilford Press, 757–76.
Rozin, P., Millman, L., and Nemeroff, C. (1986) 'Operation of the Laws of Sympathetic Magic in Disgust and Other Domains', *Journal of Personality and Social Psychology* 50(4): 703–12.
Rubbo, S.D. (1968) 'The Lethal Knife—Chemical and Biological Warfare', *The Australian Quarterly* 40(4): 21–8.
Rusk, D. (1962) 'Memorandum from the Secretary of State to the President', *Foreign Relations of the United States, 1961–1963, Volume II, 1962*: https://history.state.gov/historicaldocuments/frus1961-63v02/d270
Safire, W. (1998) 'Weapons of Mass Destruction', *New York Times*, 19 April: 22.
Sagan, S.D. and Valentino, B. (2021) 'Taboo or Not Taboo: A Rebuttal and Research Agenda', *International Studies Review* 23(3): 1092–6.
Said, E.W. (2003) *Orientalism*. London: Penguin.
Salter, M. (2014) 'Toys for the Boys? Pleasure and Popular Culture in the Militarisation of Policing', *Critical Criminology* 22: 163–77.
Sandholtz, W. (2008) 'Dynamics of International Norm Change: Rules against Wartime Plunder', *European Journal of International Relations* 14(1): 101–31.
Sanford, S., Polzer, J., and McDonough, P. (2016) 'Preparedness as a Technology of (In)Security: Pandemic Influenza Planning and the Global Biopolitics of Emerging Infectious Disease', *Social Theory & Health* 14(1): 18–43.
Sartori, A.E. (2002) 'The Might of the Pen: A Reputational Theory of Communication in International Disputes', *International Organization* 56(1): 121–49.
Sasley, B.E. (2010) 'Affective Attachments and Foreign Policy: Israel and the 1993 Oslo Accords', *European Journal of International Relations* 16(4): 687–709.
Savage, R. (2007) '"Disease Incarnate": Biopolitical Discourse and Genocidal Dehumanisation in the Age of Modernity.' *Journal of Historical Sociology* 20(3): 404–40.
Scarrow, G.D. (1988) 'The Athenian Plague: A Possible Diagnosis', *The Ancient History Bulletin* 2(1): 4–8.
Schaller, M. and Park, J.H. (2011) 'The Behavioral Immune System (and Why It Matters)', *Current Directions in Psychological Science* 20(2): 99–103.
Scheff, T. (2000) 'Shame and the Social Bond: A Sociological Theory', *Sociological Theory* 18(1): 84–99.
Scheff, T.J. (2003) 'Shame in Self and Society', *Symbolic Interaction* 26(2): 239–62.

Scheff, T.J. and Retizinger, S.M. (2002) *Emotions and Violence: Shame and Rage in Destructive Conflicts*. Lexington: Lexington Books.

Schelling, T.C. (2008) *Arms and Influence*. New Haven, CT: Yale University Press.

Schimmelfennig, F. (2000) 'International Socialization in the New Europe: Rational Action in an International Environment', *European Journal of International Relations* 6(1): 109–39.

Schmidt, A. and Sikkink, K. (2019) 'Breaking the Ban? The Heterogenous Impact of US Contestation of the Torture Norm', *Journal of Global Security Studies* 41(1): 105–22.

Schulte, P. (1999) 'Chemical and Biological Weapons: Issues and Alternatives', *Comparative Strategy* 18(4): 329–34.

Schwartz, H. (1952) 'Soviet Hate Campaign Stirs Wide Speculation', *New York Times*, 6 July: E7.

Scott, J.B. (2000a) *The Proceedings of the Hague Peace Conferences: The Conference of 1899*. New York: William S. Hein & Co.

Scott, J.B. (2000b) *The Proceedings of the Hague Peace Conferences: The Conference of 1907, Volume I*. New York: William S. Hein & Co.

Scott, R. (1965) 'Vietnam gas "propaganda gift to Communists"', *The Guardian*, 25 March: 11.

Seeley, T.D., et al. (1985) 'Yellow Rain', *Scientific American* 253(3): 128–37.

Shakespeare, F.J. (1970) 'Memorandum from the Director of the U.S. Information Agency (Shakespeare) to the President's Assistant for National Security Affairs (Kissinger)', *Foreign Relations of the United States, 1969–176, Volume E-2, Documents on Arms Control and Nonproliferation, 1969–1972*: https://history.state.gov/historicaldocuments/frus1969-76ve02/d182

Shamai, P. (2015) 'Name and Shame: Unravelling the Stigmatization of Weapons of Mass Destruction', *Contemporary Security Policy* 36(1): 104–22.

Shannon, V.P. (2000) 'Norms Are What States Make of Them: The Political Psychology of Norm Violation', *International Studies Quarterly* 44(2): 293–316.

Shapley, D. (1975) 'Impact of Protocol Passage Diluted by Senate Compromise', *Science* 187(4172): 149.

Shoham, D. and Wolfson, Z. (2004) 'The Russian Biological Weapons Program: Vanished or Disappeared?', *Critical Reviews in Microbiology* 30(4): 241–61.

Shouse, E. (2005) 'Feeling, Emotion, Affect', *M/C Journal* 8(6): http://journal.media-culture.org.au/0512/03-shouse.php

Shpiro, S. (2009) 'Poisoned Chalice: Intelligence Use of Chemical and Biological Weapons', *International Journal of Intelligence and CounterIntelligence* 22(1): 1–30.

Shultz, G.P. (1986a) 'Telegram from the Department of State to All African Diplomatic Posts', *Foreign Relations of the United States, 1981–1988, Volume XLI, Global Issues II*: https://history.state.gov/historicaldocuments/frus1981-88v41/d19

Shultz, G.P. (1986b) 'Telegram from the Department of State to All Diplomatic and Consular Posts', *Foreign Relations of the United States, 1981–1988, Volume XLI, Global Issues II*: https://history.state.gov/historicaldocuments/frus1981-88v41/d21

Shuster, A. (1969) 'Capitals in Western Europe Welcome Nixon's Move', *New York Times*, 26 November: 17.

Sikkink, K. (2013) 'The United States and Torture', in Risse, T., Ropp, S.C., and Sikkink, K. (eds), *The Persistent Power of Human Rights: From Commitment to Compliance*. Cambridge: Cambridge University Press: 145–63.

Silva, S. (2013) 'Reification and Fetishism: Processes of Transformation', *Theory, Culture & Society* 30(1): 79–98.

Simmons, J.S. (1941) 'Memorandum to H.H. Bundy', Record Group 112 (295a, Box 6), *National Archives*.

Sims, N. (2001) *The Evolution of Biological Disarmament*. Oxford: Oxford University Press.

Sims, N. (2007) 'The Future of Biological Disarmament: New Hope after the Sixth Review Conference of the Biological Weapons Convention', *Nonproliferation Review* 14(2): 351–72.
Sims, N. and Littlewood, J. (2011) 'Ambitious Incrementalism', *Nonproliferation Review* 18(3): 499–511.
SIPRI (1971a) *The Problem of Chemical and Biological Warfare V: The Prevention of CBW*. Stockholm: Almqvist & Wiksell.
SIPRI (1971b) *The Problem of Chemical and Biological Warfare Volume I: The Rise of CB Weapons*. Stockholm: Almqvist & Wiksell.
SIPRI (1971c) *The Problem of Chemical and Biological Warfare Volume VI: CB Disarmament Negotiations 1920–1970*. Stockholm: Almqvist & Wiksell.
SIPRI (1973) *The Problem of Chemical and Biological Warfare Volume III: CBW and the Law of War*. Stockholm: Almqvist & Wiksell.
Sisco, J.J. (1967) 'The UN: An Arena for Peaceful East–West Engagement', *Department of State Bulletin* LVI(1447): 458–63.
Sluka, J. (2008) 'Terrorism and Taboo: An Anthropological Perspective on Political Violence against Citizens', *Critical Studies on Terrorism* 1(2): 167–83.
Smart, J.K. (1997) 'History of Chemical and Biological Warfare: An American Perspective,' in USDoA (ed.), *Medical Aspects of Chemical and Biological Warfare*. Washington, DC: Office of the Surgeon General, 9–86.
Smetana, M. (2019) *Nuclear Deviance: Stigma Politics and the Rules of the Nonproliferation Game*. Basingstoke: Palgrave Macmillan.
Smetana, M. and O'Mahoney, J. (2022) 'NPT as an Antifragile System: How Contestations Improves the Nonproliferation Regime', *Contemporary Security Policy* 43(1): 24–49.
Smith, B. (1995) 'Gender and the Practices of Scientific History: The Seminar and Archival Research in the Nineteenth Century', *The American Historical Review* 100(4): 1150–76.
Smith, E. (1984) 'International Regulation of Chemical and Biological Weapons: Yellow Rain and Arms Control', *University of Illinois Law Review* 4: 1011–74.
Smith, F. (2014) *American Biodefense: How Dangerous Ideas about Biological Weapons Shape National Security*. Ithaca, NY: Cornell University Press.
Smith, G.C. (1969) 'Memorandum from the Director of the Arms Control and Disarmament Agency (Smith) to the President's Assistant for National Security Affairs (Kissinger)', *Foreign Relations of the United States, 1969, Volume E-2, Documents on Arms Control and Proliferation, 1969–1972*: https://history.state.gov/historicaldocuments/frus1969-76ve02/d65
Smith, G.C. (1971) 'Memorandum from the Director of the Arms Control and Disarmament Agency (Smith) to the President's Assistant for National Security Affairs (Kissinger)', *Foreign Relations of the United States, 1969–176, Volume E-2, Documents on Arms Control and Nonproliferation, 1969–1972*: https://history.state.gov/historicaldocuments/frus1969-76ve02/d252
Smith, R. (1969a) 'Laird Asks Halt in Germ Output for Use in War', *New York Times*, 18 October: 1, 7.
Smith, R. (1969b) '2 Agencies Clash over War Toxins', *New York Times*, 16 December: 1, 15.
Smith, R. (1969c) 'U.S. Debating a Cutback in Germ War Stockpile', *New York Times*, 24 September: 4.
Smith, R.A. and Hughes, D. (2014) 'Infectious Disease Stigmas: Maladaptive in Modern Society', *Communication Studies* 65(2): 132–8.
Smith, S. (2001) 'Globalization and the Governance of Space: A Critique of Krasner on Sovereignty', *International Relations of the Asia-Pacific* 1(2): 199–226.
Smithson, A.E. (2004) 'Biological Weapons: Can Fear Overwhelm Inaction?', *The Washington Quarterly* 28(1): 165–78.
Sobel, J. (1985) 'A Theory of Credibility', *The Review of Economic Studies* 52(4): 557–73.

Sontag, S. (1978) *Illness as Metaphor*. New York: Farrar, Straus and Giroux.
Sophocles (2008) *Electra and Other Plays [Women in Trachis]*. Translated by Raeburn, D. London: Penguin.
Souers, S.W. (1948) 'Report to the President by the National Security Council', *Foreign Relations of the United States, 1948, General; The United Nations, Volume I, Part 2*: 662–9.
Spiers, E.M. (2010) *A History of Chemical and Biological Weapons*. London: Reaktion.
Sprague, C.A. (1952) 'Performance of the Dependent Press in the Soviet Union', *Department of State Bulletin* XXVII(702): 920–3.
Stafford, M.C. and Scott, R.R. (1986) 'Stigma Deviance and Social Control: Some Conceptual Issues', in Ainlay, S.C., Becker, G., and Coleman, L.M. (eds), *The Dilemma of Difference*. New York: Plenum, 77–91.
Steele, B. (2008) *Ontological Security in International Relations: Self-Identity and the IR State*. London: Routledge.
Steelman, J.R. (1949) 'The Chairman of the National Security Resources Board (Steelman) to the Secretary of State', *Foreign Relations of the United States, 1949, National Security Affairs, Foreign Economic Policy, Volume I*: 336–45.
Stein, E. (1972) 'Legal Restraints in Modern Arms Control Agreements', *The American Journal of International Law* 66(2): 255–89.
Steinbruner, J. (1997) 'Biological Weapons: A Plague upon All Houses', *Foreign Policy* 109: 85–96.
Steinbruner, J. and Okutani, S. (2004) 'The Protective Oversight of Biotechnology', *Biosecurity and Bioterrorism: Biodefense Strategy, Practice and Science* 2(4): 273–80.
Steiner, F. (1956) *Taboo*. London: Routledge.
Stern, J. (1999) *The Ultimate Terrorists*. Cambridge, MA: Harvard University Press.
Stern, J. (2002) 'Dreaded Risks and the Control of Biological Weapons', *International Security*, 27(3): 89–123.
Stevens, A. (1950) 'U.S. Pushes Study of Germ Warfare', *New York Times*, 23 April: 78.
Stevenson, E. (1950) *Report of the Secretary of Defense's Ad Hoc Committee on Chemical, Biological and Radiological Warfare*.
Stimson, H. (1942) 'Henry L. Stimson to President, April 29', File 104, *Roosevelt Papers*.
Stossel, W.J. (1981) 'Reported Use of Chemical Weapons', *Department of State Bulletin* 81(2056): 79.
Stossel, W.J. (1982) 'U.S. Issues Report on Chemical Warfare', *Department of State Bulletin* 82(2062): 57.
Strohminger, N. (2014) 'Disgust Talked About', *Philosophy Compass* 9(7): 478–93.
Synge, A. (1969) 'Chemical and Biological Warfare', *New Blackfriars* 50(585): 256–63.
Szabados, B. (1979) 'Hypocrisy', *Canadian Journal of Philosophy* 9(2): 195–210.
Tannenwald, N. (1999) 'The Nuclear Taboo: The United States and the Normative Basis of Nuclear Non-Use', *International Organization* 53(3): 433–68.
Tannenwald, N. (2007) *The Nuclear Taboo: The United States and the Non-Use of Nuclear Weapons Since 1945*. Cambridge: Cambridge University Press.
Tatu, L., Jost, W., and Bogousslavsky, J. (2017) 'The Botulinum Toxin Legend of Reinhard Heydrich's Death', *Historical Neurology* 89(1); 84–7.
Tetlock, P.E. (2003) 'Thinking the Unthinkable: Sacred Values and Taboo Cognitions', *TRENDS in Cognitive Sciences* 7(7): 320–4.
Thomas, W. (2000) 'Norms and Security: The Case of International Assassination', *International Security* 25(1): 105–33.
Thucydides (2004) *The History of the Peloponnesian War*. Translated by Crawley, R. Project Guttenberg: https://www.gutenberg.org/files/7142/7142-h/7142-h.htm

Tingley, D.H. and Walter, B.F. (2011) 'The Effect of Repeated Play on Reputation Building: An Experimental Approach', *International Organization* 65(2): 343–65.

Tomes, N. (2002) 'Epidemic Entertainments: Disease and Popular Culture in Early-Twentieth-Century America', *American Literary History* 14(4): 625–52.

Tomz, M. (2007) *Reputation and International Cooperation: Sovereign Debt across Three Centuries*. Princeton, NJ: Princeton University Press.

Toros, H. (2008) '"We Don't Negotiate with Terrorists!": Legitimacy and Complexity in Terrorist Conflicts', *Security Dialogue* 38(4): 407–26.

Toy, C.H. (1899) 'Taboo and Morality', *Journal of the American Oriental Society* 20: 151–6.

Trent, W. (1924) 'William Trent's Journal at Fort Pitt, 1763', *The Mississippi Valley Historical Review* 11(3): 390–413.

Truman, H.S. (1948) 'Address in Miami at the American Legion Convention', *Truman Presidential Library*: https://www.trumanlibrary.gov/library/public-papers/244/address-miami-american-legion-convention

Truman, H.S. (1952) 'Remarks to Members of the American Action Committee Against Mass Deportations in Romania', *Truman Presidential Library*: https://www.trumanlibrary.gov/library/public-papers/145/remarks-members-american-action-committee-against-mass-deportations

Truman, H.S. (1953) 'Letter from Truman to Murray', Box 112, *Truman Papers*.

Truman, H.S. (1955) *1945 Year of Decisions: Memoirs by Harry S. Truman, Volume I*. New York: Time.

Tsutsui, W.M. (2007) 'Looking Straight at "Them!": Understanding the Big Bug Movies of the 1950s', *Environmental History* 12(2): 237–53.

Tucker, J.B. (1995) 'Strengthening the Biological Weapons Convention', *Arms Control Today* 25(3): 9–12.

Tucker, J.B. (1997) 'The Biological Weapons Threat', *Current History* 96(609): 167–72.

Tucker, J.B. (2000) 'Motivations for and against Proliferation: The Case of the Middle East', in Zilinskas, R.A. (ed.), *Biological Warfare: Modern Offence and Defense*. London: Lynne Reiner, 27–52.

Tucker, J.B. (2001a) 'Lessons from the Case Studies', in J. B. Tucker (ed.), *Toxic Terror: Assessing Terrorist Use of Chemical and Biological Weapons*, Cambridge, MA: MIT Press, 249–70.

Tucker, J.B. (2001b) 'The "Yellow Rain" Controversy: Lessons for Arms Control Compliance', *Nonproliferation Review* 8(1): 25–42.

Tucker, J.B. (2002) 'A Farewell to Germs: The U.S. Renunciation of Biological and Toxin Warfare', *International Security* 27(1): 107–48.

Tucker, J.B. (2010) 'Seeking Biosecurity without Verification: The New U.S. Strategy on Biothreats', *Arms Control Today* 40(1): 8–14.

Ulm, A.H. (1926) 'Senate to Debate Warfare with Gas', *New York Times*, 31 October: 5.

UN (1948) *Resolution of the Commission for Conventional Armaments: Definition of Armaments, August 12, 1948.*

UN (1966) *Resolution 2162 (XXI) B: Question of general and complete disarmament.*

UN (1969a) 'Report by the Secretary-General Thant on Chemical and Bacteriological (Biological) Weapons and the Effect of Their Possible Use [Extracts]', *Documents on Disarmament 1969*: 264–98.

UN (1969b) *Resolution 2603 (XXIV): Question of chemical and bacteriological (biological) weapons.*

US (2009) *National Strategy for Countering Biological Threats.*

US (2018) *National Biodefense Strategy.*

USCong (1959) *Chemical, Biological, and Radiological Warfare Agents: Hearings Before the Committee on Science and Astronautics, U.S. House of Representatives, Eighty-Sixth Congress, First Session*. Washington, DC: U.S. Government Printing Office.

USCong (1960) *Chemical-Biological-Radiological (CBR) Warfare and Its Disarmament: A Study Prepared by the Subcommittee on Disarmament of the Committee on Foreign Relations, United States Senate*. Washington, DC: U.S. Government Printing Office.

USCong (1969a) *Chemical-Biological Warfare: U.S. Policies and International Effects: Hearings Before the Subcommittee on National Security Policy and Scientific Developments of the Committee on Foreign Affairs, House of Representatives, Ninety-First Congress, First Session*. Washington, DC: U.S. Government Printing Office.

USCong (1969b) *Environmental Dangers of Open Testing of Lethal Chemicals: Hearings Before a Subcommittee on Government Operations, House of Representatives, Ninety-First Congress, First Session*. Washington, DC: U.S. Government Printing Office.

USCong (1969c) *International Implications of Dumping Poisonous Gas and Waste Into Oceans: Hearings Before the Subcommittee on International Organizations and Movements of the Committee on Foreign Affairs, House of Representatives, Ninety-First Congress, First Session*. Washington, DC: U.S. Government Printing Office.

USCong (1970) *Chemical-Biological Warfare: U.S. Policies and International Effects: Report of the Subcommittee on National Security and Scientific Developments of the Committee on Foreign Affairs, House of Representatives*. Washington, DC: U.S. Government Printing Office.

USCong (1971) *The Geneva Protocol of 1925: Hearing before the Committee on Foreign Relations, United States Senate, Ninety-Second Congress, First Session*. Washington, DC: U.S. Government Printing Office.

USCong (1974a) *Prohibition of Chemical and Biological Weapons: Hearing before the Committee on Foreign Relations, United States Senate, Ninety-Third Congress, Second Session*. Washington, DC: US Printing Office.

USCong (1974b) 'Report by the House Foreign Affairs Committee on Ratification of the Geneva Protocol', *Documents on Disarmament 1974*: 369–79.

USCong (1975) *Hearings before the Select Committee to Study Governmental Operations With Respect to Intelligence Activities of the United States Senate, Ninety-Fourth Congress, First Session, Volume 1: Unauthorized Storage of Toxic Agents*. Washington, DC: US Government Printing Office.

USCong (1976) *Final Report of the Select Committee to Study Government Operations with Respect to Intelligence Activities, United States Senate Together with Additional, Supplemental, and Separate Views*. Washington, DC: US Printing Office.

USCong (1977) *Project MKULTRA, the CIA's Program of Research in Behavioral Modification: Joint Hearing before the Select Committee on Intelligence and the Subcommittee on Health and Scientific Resources, United States Senate, Ninety-Fifth Congress, First Session, August 3, 1977*. Washington, DC: US Printing Office.

USSR (1983) 'Letter from the Soviet Union to U.N. Secretary-General Perez de Cuellar: U.S. Report on Chemical Warfare in Southeast Asia and Afghanistan', *Documents on Disarmament 1983*: 79–80.

van Atta, D. (2008) *With Honor: Melvin Laird in War, Peace, and Politics*. Madison: University of Wisconsin Press.

van Courtland Moon, J.E. (1999a) 'US Biological Warfare Planning and Preparedness: The Dilemmas of Policy', in Geissler, E. and van Courtland Moon, J.E. (eds), *Biological and Toxin Weapons: Research, Development and Use from the Middle Ages to 1945*. Oxford: Oxford University Press: 215–54.

van Courtland Moon, J.E. (1999b) 'Dubious Allegations', *The Bulletin of the Atomic Scientists* 55(3): 70–2.

van Courtland Moon, J.E. (2009) 'The US Biological Weapons Program', in Wheelis, M., Roizsa, L., and Dando, M. (eds), *Deadly Cultures: Biological Weapons since 1945*. Cambridge, MA: Harvard University Press: 9–46.

van Ginneken, J. (1977) 'Bacteriological Warfare', *Journal of Contemporary Asia* 7(2): 130–52.

van Ham, P. (2001) 'The Rise of the Brand State: The Postmodern Politics of Image and Reputation', *Foreign Affairs* 80(5): 2–6.

Vance, C. (1980) 'Telegram from the Department of State to the Embassy in the Soviet Union and the U.S. Mission at Geneva', *Foreign Relations of the United States, 1977–1980, Volume VI, Soviet Union*: https://history.state.gov/historicaldocuments/frus1977-80v06/d267

Vermeij, L. (2020) *Woman First, Solider Second: Taboos and Stigma Facing Military Women in UN Peace Operations*. New York: International Peace Institute.

Vestergaard, C. and Roul, A. (2011) 'A (F)utile Intersessional Process?', *Nonproliferation Review* 18(3): 489–97.

Waggoner, W.H. (1949) 'Forrestal Scouts Germ Warfare as Army Says U.S. is Ready', *New York Times*, 13 March: 1, 36.

Waggoner, W.H. (1952a) 'Peiping First Held Epidemics Natural', *New York Times*, 20 April: 1, 3.

Waggoner, W.H. (1952b) 'Experts Discount Amity from Soviet', *New York Times*, 19 June: 3.

Walker, W. (2010) 'The Absence of a Taboo on the Possession of Nuclear Weapons', *Review of International Studies* 36(4): 865–76.

Walt, S. (2013) 'Types of Weapons Assad Uses Shouldn't Affect U.S. Policy', *New York Times*, 26 August: https://www.nytimes.com/roomfordebate/2013/08/26/is-an-attack-on-syria-justified/type-of-weapons-assad-uses-shouldnt-affect-us-policy

Ward, K.D. (2004) 'The BWC Protocol: Mandate for Failure', *Nonproliferation Review* 11(2): 183–99.

Wark, S. and Wark, M. (2019) 'Circulation and its Discontents', in Bristow, D. and Brown, A. (eds), *Post Memes: Seizing the Memes of Production*. Goleta: Punctum Books, 293–318.

Weathersby, K. (1998) 'Deceiving the Deceivers: Moscow, Beijing, Pyongyang, and the Allegations of Bacteriological Weapons Use in Korea', *Cold War International History Bulletin* 11: 176–85.

Weaver, W. (1969) 'Laird Backs Senate Curb on Chemical War', *New York Times*, 10 August: 1, 29.

Webster, H. (1942) *Taboo: A Sociological Study*. Stanford, CA: Stanford University Press.

Webster, W.H. (1988) 'Remarks at the World Affairs Council of Washington D.C., October 25': https://www.cia.gov/library/readingroom/docs/CIA-RDP99-00777R000302380002-7.pdf

Weinberger, C.A. (1983) 'Report by Secretary of Defense Weinberger to the Congress: The Role of Nuclear Weapons in U.S. Strategy and Deterrence of Chemical Warfare [Extracts], February 1, 1983', *Documents on Disarmament 1983*: 52–64.

Weiss, M.G., Ramakrishna, J., and Somma, D. (2006) 'Health-Related Stigma: Rethinking Concepts and Interventions', *Psychology, Health & Medicine* 11(3): 277–87.

Wendt, A. (1999) *Social Theory of International Politics*. Cambridge: Cambridge University Press.

Wessely, S., Hyams, K.C., and Bartholomew, R. (2001) 'Psychological Implications of Chemical and Biological Weapons: Long Term Social and Psychological Effects May Be Worse than Acute Ones', *BMJ* 323(7318): 878–9.

Westing, A.H. (1985a) 'The Threat of Biological Warfare', *BioScience* 35(10): 627–33.

Westing, A.H. (1985b) 'Towards Eliminating the Scourge of Chemical War: Reflections on the Occasion of the Sixtieth Anniversary of the Geneva Protocol', *Bulletin of Peace Proposals* 16(2): 117–20.

Westing, A.H. (1989) 'Chemical and Biological Warfare: The Road Not Taken', *Peace Research* 21(1): 17–20, 81–3.

Wheelis, M. (1999a) 'Biological Warfare before 1914', in Geissler, E. and van Courtland Moon, J.E. (eds), *Biological and Toxin Weapons: Research, Development and Use from the Middle Ages to 1945*. Oxford: Oxford University Press, 8–34.

Wheelis, M. (1999b) 'Biological Sabotage in World War I', in Geissler, E. and van Courtland Moon, J.E. (eds), *Biological and Toxin Weapons: Research, Development and Use from the Middle Ages to 1945*. Oxford: Oxford University Press, 35–62.

Wheelis, M. (2004) 'Will the "New Biology" Lead to New Weapons?', *Arms Control Today* 34(6): 6–13.

White, A.D. (1905) *Autobiography of Andrew Dickson White: Volume II*. Project Gutenberg: http://www.gutenberg.org/ebooks/1370

White House (1970) 'U.S. Renounces Use of Toxins as a Method of Warfare', *Department of State Bulletin* LXII(1601): 226–7.

WHO (2004) *Public Health Response to Biological and Chemical Weapons: WHO Guidance*. Second edition. Geneva: World Health Organization.

Wickham-Steed, H. (1934) 'Aerial Warfare: Secret German Plans', *Nineteenth Century and After* 116: 1–15.

Williamson, D. (2012) *The 47th Indiana Volunteer Infantry: A Civil War History*. Jefferson, NC: McFarland & Company.

Winchester, S. (1975) 'CIA chiefs "ignored order by Nixon" to destroy poisons', *The Guardian*, 17 September: 2.

Witt, J.F. (2014) 'Civil War Historians and the Laws of War', *Journal of the Civil War Era* 4(2): 159–71.

Wolford, S. (2007) 'The Turnover Trap: New Leaders, Reputation, and International Conflict', *American Journal of Political Science* 51(4): 772–88.

Wong, A.S. (2016) 'Emotions and the Communication of Intentions in Face-to-Face Diplomacy', *European Journal of International Relations* 22(1): 144–67.

WPC (1952) *Report of the International Scientific Commission for the Investigation of the Facts Concerning Bacterial Warfare in Korea and China, 1952*.

Wylie, J.A.H. and Stubbs, H.W. (1983) 'The Plague of Athens: 430-428 B.C. Epidemic and Epizootic', *The Classical Quarterly* 33(1): 6–11.

Yost, C.W. (1970) 'Telegram 122 From the Mission to the United Nations to the Department of State', *Foreign Relations of the United States, 1969–176, Volume E-2, Documents on Arms Control and Nonproliferation, 1969–1972*: https://history.state.gov/historicaldocuments/frus1969-76ve02/d175

Young, K. (1928) 'A Story of the Rise of a Social Taboo', *The Scientific Monthly* 26(5): 449–53.

Zanders, J.P. (1999) 'Assessing the Risk of Chemical and Biological Weapons Proliferation to Terrorists', *Nonproliferation Review* 6(4): 17–34.

Zanders, J.P. (2001) 'Challenges to Disarmament Regimes: The Case of the Biological and Toxin Weapons Convention', *Global Society* 15(4): 361–85.

Zanders, J.P. (2003) 'International Norms against Chemical and Biological Warfare: An Ambiguous Legacy', *Journal of Conflict & Security Law* 8(2): 391–410.

Zarakol. A. (2011) *After Defeat: How the East Learned to Live with the West*. Cambridge: Cambridge University Press.

Zerka, P. (2021) 'Why Should Anyone Care? Foreign Policy and Public Opinion', European Council on Foreign Relations, 19 April: https://ecfr.eu/article/why-should-anyone-care-foreign-policy-and-public-opinion/

Zilinskas, R.A. (2016) *The Soviet Biological Weapons Program and Its Legacy in Today's Russia, Occasional Paper 11*. Washington, DC: National Defense University Press.

Zizek, S. (1989) *The Sublime Object of Ideology*. London: Verso.

Zuesse, E.M. (1974) 'Taboo and the Divine Order', *Journal of the American Academy of Religion* 42(3): 482–504.

Index

For the benefit of digital users, indexed terms that span two pages (e.g., 52–53) may, on occasion, appear on only one of those pages.